Brett Callaway

Compendium of Corn Diseases

SECOND EDITION

Edited by
Malcolm C. Shurtleff

Published by The American Phytopathological Society

In cooperation with:

The Cooperative Extension Service and the
Department of Plant Pathology, University of Illinois, Urbana

APS PRESS
The American Phytopathological Society

Editor:
Malcolm C. Shurtleff, University of Illinois, Urbana, IL

Coordinators:
T. L. Whatley, University of Illinois, Urbana, IL
Thor Kommedahl, University of Minnesota, St. Paul, MN

Cooperators and Advisory Committee:
S. R. Aldrich, University of Illinois, Urbana, IL
D. C. Arny, University of Wisconsin, Madison, WI
K. R. Barker, North Carolina State University, Raleigh, NC
R. R. Bergquist, Pfister Hybrid Corn Company, El Paso, IL
S. S. Bhat, University of Mysore, India
L. M. Black, University of Illinois, Urbana, IL
C. W. Boothroyd, Cornell University, Ithaca, NY
C. M. Christensen, University of Minnesota, St. Paul, MN
L. J. Claflin, Kansas State University, Manhattan, KS
C. J. D'Arcy, University of Illinois, Urbana, IL
V. D. Damsteegt, USDA-SEA, Agricultural Research, Frederick, MD
R. E. Davis, USDA-SEA, Agricultural Research, Beltsville, MD
Carlos DeLeon, Maize Program CIMMYT, Mexico City, Mexico
U. L. Diener, Auburn University, Auburn, AL
J. L. Dodd, Cargill, Inc., Aurora, IL
Ben Doupnik, Jr., University of Nebraska, Clay Center, NE
J. P. Downs, University of Illinois, Urbana, IL
H. E. Duncan, North Carolina State University, Raleigh, NC
B. Y. Endo, USDA-SEA, Agricultural Research, Beltsville, MD
A. H. Epstein, Iowa State University, Ames, IA
H. S. Fenwick, University of Idaho, Moscow, ID
V. R. and J. M. Ferris, Purdue University, W. Lafayette, IN
R. E. Ford, University of Illinois, Urbana, IL
R. A. Frederiksen, Texas A & M University, College Station, TX
W. N. Garrett, University of Georgia, Athens, GA
D. T. Gordon, Ohio Agricultural Research & Development Center, Wooster, OH
R. G. Hoeft, University of Illinois, Urbana, IL
Quintin Holdeman, California Department of Agriculture, Sacramento, CA
A. L. Hooker, University of Illinois, Urbana, IL
C. W. Horne, Texas A & M University, College Station, TX
R. S. Hussey, University of Georgia, Athens, GA
B. J. Jacobsen, University of Illinois, Urbana, IL
A. W. Johnson, USDA-SEA, Agricultural Research, Tifton, GA
R. W. Jugenheimer, University of Illinois, Urbana, IL
A. L. Karr, University of Missouri, Columbia, MO
Arthur Kelman, University of Wisconsin, Madison, WI
R. G. Kenneth, The Hebrew University of Jerusalem, Rehovot, Israel
E. L. Knake, University of Illinois, Urbana, IL
L. C. Lane, University of Nebraska, Lincoln, NE
S. M. Lim, University of Illinois, Urbana, IL
Frances M. Latterell, USDA-SEA, Agricultural Research, Frederick, MD
E. S. Luttrell, University of Georgia, Athens, GA
W. F. Mai, Cornell University, Ithaca, NY
R. B. Malek, University of Illinois, Urbana, IL
Karl Maramorosch, Rutgers University, New Brunswick, NJ
M. C. McGlamery, University of Illinois, Urbana, IL
C. J. Mirocha, University of Minnesota, St. Paul, MN
L. R. Nault, Ohio Agricultural Research & Development Center, Wooster, OH
R. R. Nelson, Pennsylvania State University, University Park, PA
C. L. Niblett, Kansas State University, Manhattan, KS
D. C. Norton, Iowa State University, Ames, IA
B. E. Nowlin, University of Illinois, Urbana, IL
E. W. Palm, University of Missouri, Columbia, MO
V. G. Perry, University of Florida, Gainesville, FL
B. L. Renfro, The Rockefeller Foundation, Bangkok, Thailand
S. M. Ries, University of Illinois, Urbana, IL
C. C. Russell, Oklahoma State University, Stillwater, OK
P. F. Sand, USDA-APHIS, Hyattsville, MD
N. W. Schaad, University of Georgia, Experiment, GA
M. L. Schuster, University of Nebraska, Lincoln, NE
W. O. Scott, University of Illinois, Urbana, IL
C. G. Shaw, Washington State University, Pullman, WA
Odette L. Shotwell, USDA-SEA, Northern Regional Research Center, Peoria, IL
F. L. Slife, University of Illinois, Urbana, IL
D. R. Smith, DeKalb Ag. Research, Inc., DeKalb, IL
R. E. Stuckey, University of Kentucky, Lexington, KY
D. R. Sumner, University of Georgia, Tifton, GA
John Tuite, Purdue University, W. Lafayette, IN
M. T. Turner, Funk Seeds International, Inc., Bloomington, IL
F. A. Uecker, USDA Mycology Laboratory, Beltsville, MD
A. J. Ullstrup, FFR Cooperative, W. Lafayette, IN
J. K. Uyemoto, Kansas State University, Manhattan, KS
Anne K. Vidaver, University of Nebraska, Lincoln, NE
H. L. Warren, USDA-SEA, Purdue University, W. Lafayette, IN
D. G. White, University of Illinois, Urbana, IL
D. R. Wilkinson, Pioneer HiBred International, Inc., Johnson, IA
G. L. Worf, University of Wisconsin, Madison, WI
T. D. Wyllie, University of Missouri, Columbia, MO

Front and back covers: Alvin L. Lang, University of Illinois, Urbana

Published 1973. Second Edition 1980. Second Printing 1984. Third Printing 1986.

ISBN:0-89054-029-2
Library of Congress Catalog Card Number: 80-67517

The American Phytopathological Society
3340 Pilot Knob Road, St. Paul, Minnesota 55121

Printed in the United States of America

Contents

Preface

The purpose of this second edition of the Corn (or Maize) Compendium is to provide a thorough, authoritative, and practical reference for plant pathologists and other agriculturists engaged in diagnosing crop problems in the field or plant clinic laboratory. It is designed especially for the extension or advisory plant pathologist who has access to a microscope and for agricultural workers whose technical training in field pathology of maize or corn may be limited. Hopefully, this edition of the Compendium will prove as useful as its predecessor to crop disease consultants and their scouts, advisers in state and federal departments of agriculture and regulatory agencies, agribusiness representatives, pest control dealers and applicators, county agricultural agents, area crop specialists, educators, students in the plant sciences, growers, and researchers throughout the world.

The Compendium is amply illustrated, uses descriptive terminology, and includes totally new sections on mycoplasmal diseases, mycotoxins, and mycotoxicoses. The bacterial, viral, and nematode sections have been completely rewritten and introductory material on the major types of causal agents has been included. The glossary has been enlarged, a detailed index has been added, and there are many new photographs and selected references. The 200 references listed are a blend of the most contempory and classic scientific papers, books, monographs, and workshop proceedings currently available. More than 10,000 scientific and extension-type writings were consulted during the preparation of the manuscript.

Any abnormality of the maize plant or of its seed is considered herein to be a disease. Injuries due to insects, however, are discussed only as they relate to, resemble, or may confuse other diagnoses of disease. Over 100 infectious diseases comprise the first and largest section of the Compendium. Abiotic or noninfectious diseases (or disorders), caused by a variety of environmental factors, comprise the second section.

Frequent cross-references in the text refer the reader to similarities and contrasts between different diseases and detailed descriptions of pathogens. Symptoms, pathogens, and disease cycles are given special attention. Much of the critical history, economic importance, known distribution of diseases, and epidemiology of all major diseases is included.

Suggested control measures are purposely general, stressing principles and cultural practices that should not become obsolete and are applicable in many maize-growing areas of the world. For specific current information on resistant hybrids and varieties, chemical control, and other measures the reader should consult recent literature, competent area specialists, extension or advisory plant pathologists, or informed seed suppliers.

The diseases have been arranged according to general causal agents: bacterial diseases, mycoplasmal diseases, fungal diseases, viral diseases, nematode diseases, parasitic flowering plants (witchweed), and noninfectious or abiotic problems. The diseases and disorders, as well as illustrative materials within each group, have been placed together as they would most likely appear in the field or could be confused, e.g., leaf spots and blights, downy mildews, stalk rots, rusts, and ear and kernel rots.

A list of scientific names for hosts other than maize is provided on page 88.

Future revisions of this Compendium are planned at about five-year intervals to keep the information current and to make improvements in format, the amount and level of detail, quality of the illustrations, and other items as needed.

With your suggestions, future editions will increase in practicality and value for the diagnosis of corn (maize) diseases. Numerous suggestions provided by readers of the first edition of this Compendium have been incorporated into this edition.

The Advisory Committee and Cooperators wish to thank the following individuals who contributed illustrations and assisted in other ways. Without the help of these individuals this edition of the Compendium could not have been completed.

G. N. Agrios	H. G. Johnson
J. A. Berry	A. A. Lopez
W. A. Feder	N. E. McGlohon
A. M. Finley	J. M. Prescott
N. W. Frazier	G. S. Purss
J. H. Freitag	Saul Rich
R. L. Gamez	H. H. Thornberry
P. Grancini	R. W. Toler
I. Harpaz	M. B. von Wechmar
Frieda Herold	J. C. Wells
A. C. Hill	W. G. Willis

Mycological drawings used in this Compendium were prepared by the editor.

Introduction

Maize (corn), along with rice and wheat, are the three most important cereal crops in the world. Maize is grown throughout the temperate, subtropical, and tropical zones wherever rainfall or irrigation is adequate. The crop is grown on more than 100 million ha each year, with an annual production of about 250 million metric tons.

Corn is the most important plant native to the American continent. The United States produces about 44% of the world crop. Directly or indirectly, corn provides the USA with more food than any other cereal crop.

Other leading maize-producing countries include Argentina, Brazil, China, France, India, Indonesia, Italy, Mexico, The Philippines, Romania, South Africa, USSR, and Yugoslavia.

Although grown principally for animal feed, maize is also important as a staple food crop for human consumption, particularly in Mexico and other Central and South American countries, Africa, and Asia. Maize has more than 500 other uses; more industrial products are made from it than from any other grain.

The origin of the crop is lost in antiquity. Excavations and measures of radioactive decay of ancient ears of maize found in caves and burial sites indicate that the plant has served man for over 7,000 years. Its origin is probably in Mexico, Central America, the highlands of northern South America, or the southwestern USA.

The average world trade of maize is about 25 million metric tons per year with the USA exporting about 60% of the world market.

All parts of the maize plant are susceptible to a number of diseases that reduce the yield and quality of the crop. World reduction in grain production from diseases is estimated to average 9.4%. Annual losses in the Corn Belt of the USA range from about 7 to 17%. These diseases may be classified as infectious (parasitic or biotic) or noninfectious (nonparasitic or abiotic). Infectious diseases are caused by bacteria, mycoplasmas and spiroplasmas, fungi, viruses, nematodes, and flowering plants. Noninfectious diseases result from unfavorable climatic and soil conditions, chemical or mechanical injuries, and genetic abnormalities. Nutritional deficiencies or excesses, too much or too little water, air pollutants, or excessively high or low temperatures may produce symptoms that resemble those caused by infectious agents.

Disease Development

The extent and severity of infectious maize diseases depend on the presence of a virulent pathogen, the proper air and soil environment, and the susceptibility of the maize host. Insects or other vectors (carriers or transmitting agents) are necessary to spread some bacterial, viral, and mycoplasmalike pathogens. These factors must be present and "in balance" for an infectious disease to develop. The relationship among these conditions is shown in Fig. 1.

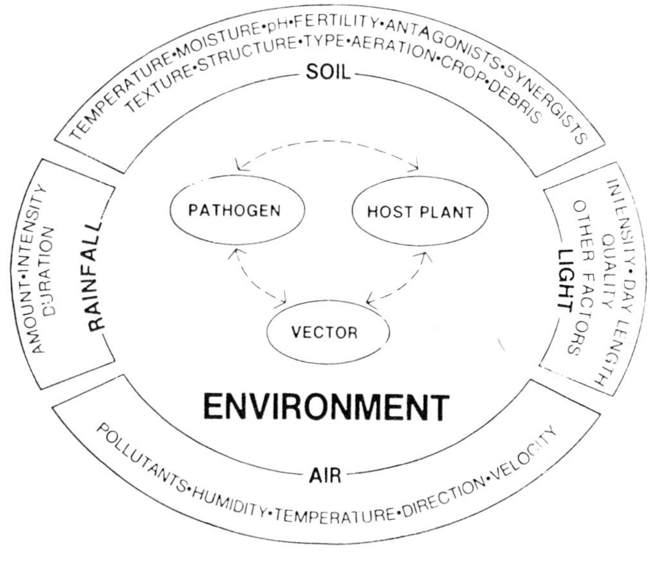

Fig. 1. Environmental influences on infectious plant diseases.

Infectious Diseases

Fungi

Fungi are simple, filamentous plants without chlorophyll. Typically, they persist, reproduce, and spread by spores that germinate under favorable conditions of temperature and moisture to produce branched threads called hyphae. An aggregate of hyphae forms the fungus body or mycelium. Fungi infect plants by direct penetration of the cuticle, through natural openings or wounds. Resting bodies (chlamydospores, sclerotia) allow fungi to survive unfavorable conditions for long periods. A generalized fungal disease cycle is shown in Fig. 2.

Bacteria

Bacteria are microscopic, one-celled organisms that enter plants through wounds or small natural openings such as stomata or hydathodes. They are spread by man through many cultural practices, in plant materials (seeds, transplants, nursery stock), and by insects, splashing rain, or flowing water. Once within the host, bacteria develop intercellularly and intracellularly, migrating through the plant. Bacteria survive in living or dead plants, seeds, soil, or in insect bodies. All plant-pathogenic species are rod-shaped and do not form spores.

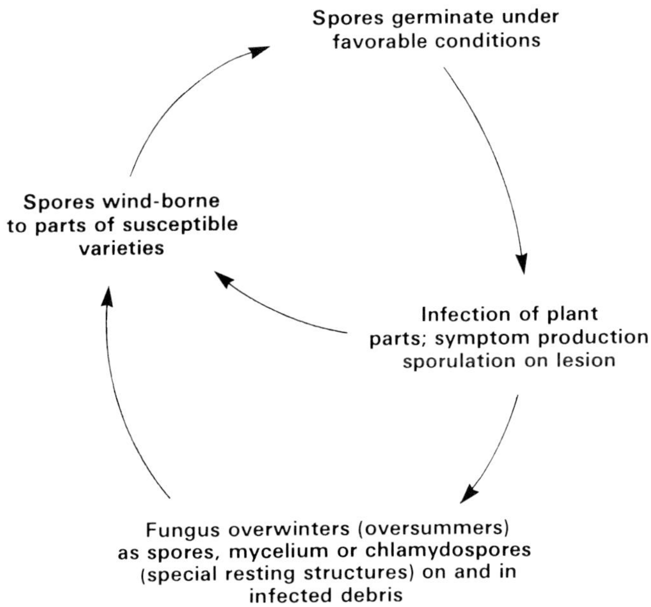

Fig. 2. Generalized life cycle of fungus.

Mycoplasmas

Mycoplasmas are minute, pleomorphic bacteria that lack a rigid cell wall. Like some viruses, they occur only in phloem cells of diseased plants and are transmitted by leafhoppers, dodder, grafting, and propagation of infected plant parts. Mycoplasmas and spiroplasmas produce viruslike symptoms in maize plants including chlorosis, stunting, phyllody, proliferation of axillary shoots, and sterility. Unlike viruses, they are sensitive to the tetracycline class of antibiotics.

Viruses

Viruses are submicroscopic, long or short rods, roughly spherical or polyhedral particles composed of nucleic acid and protein (Fig. 3). They are transmitted mechanically under experimental conditions or, in nature, by biological vectors, most importantly aphids, leafhoppers, planthoppers, beetles, and other insects. Few maize viruses are seed-transmitted. Viruses survive in weed and crop plants or in insect bodies and induce a variety of local or systemic symptoms, including mosaic patterns, yellowing or reddening of the foliage, stunting, ringspots, and necrosis.

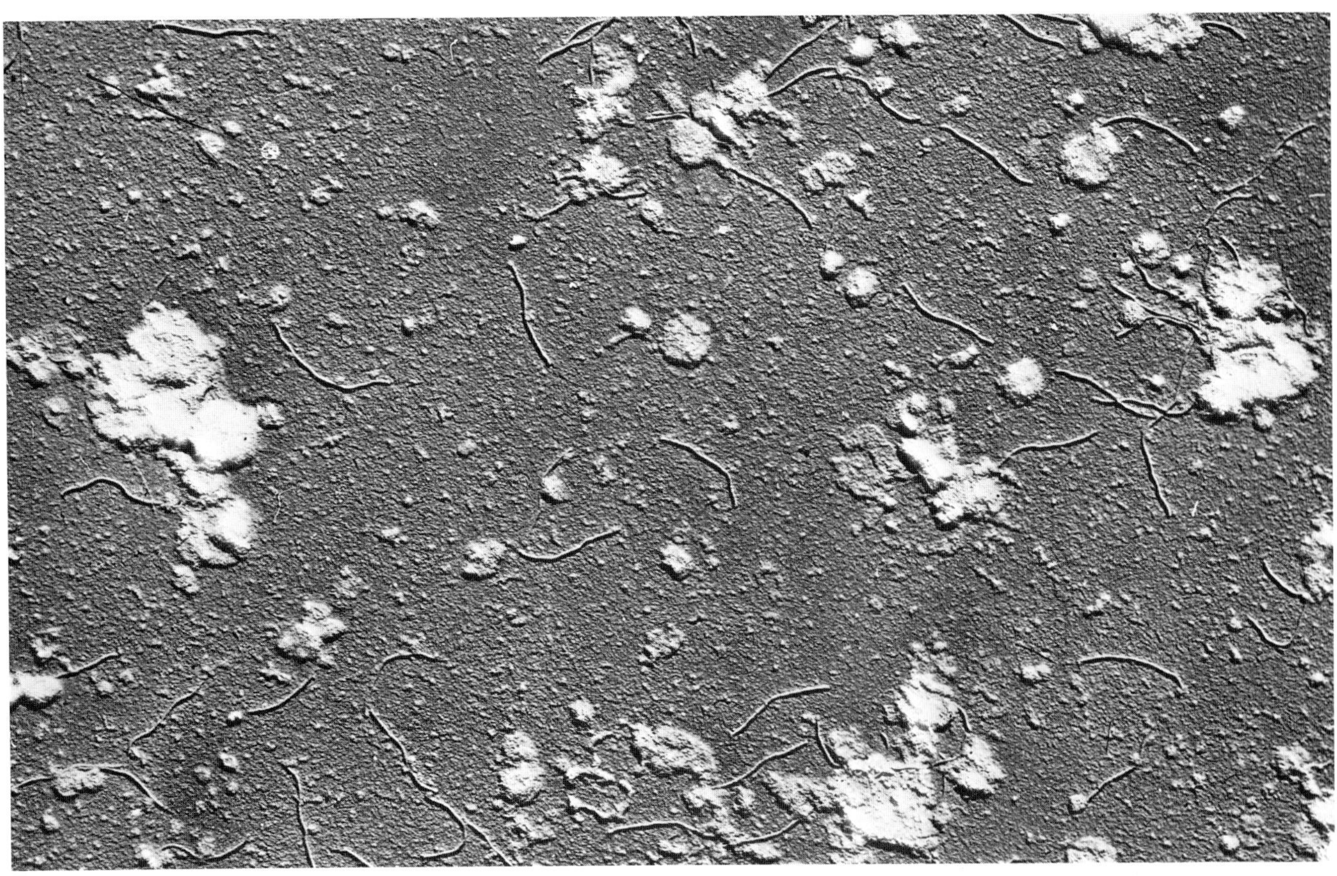

Fig. 3. Sugarcane mosaic. Electron micrograph of purified, aggregated SMCV particles after zone electrophoresis. The specimen was shadowcast with tungstic oxide. (Courtesy M. B. von Wechmar)

Nematodes

Nematodes are small, eel-shaped animals (Fig. 4). They are usually 1–2 mm long. Almost all plant-parasitic species feed by puncturing plant cells and extracting the predigested contents with a hollow spear or stylet. Nematodes are disseminated by infested soil, nursery stock, running water, or vehicles and machinery. A generalized nematode life cycle is shown in Fig. 5.

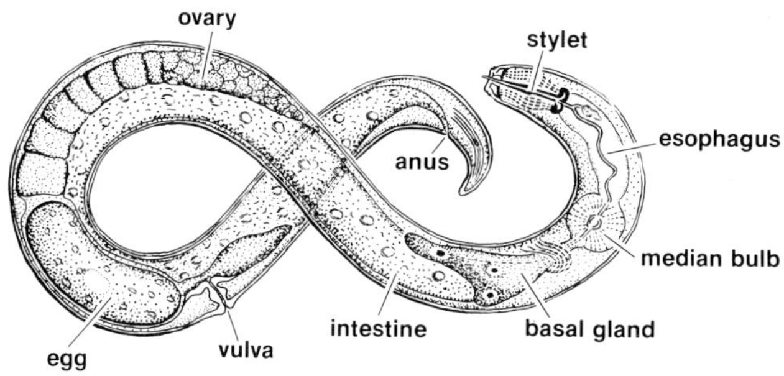

Fig. 4. Root-lesion nematode (*Pratylenchus zeae*). (Drawing by C. C. Russell, Oklahoma State University)

Parasitic Seed Plants

Parasitic seed plants, principally witchweed (*Striga* spp.), attack the roots of maize and other gramineous hosts in some subtropical and tropical areas. Plants are infected by haustoria entering the root tissues of the host (page 77). They rely mainly on seeds to spread.

Noninfectious Diseases

Noninfectious or abiotic diseases are caused by an excess, deficiency, or imbalance of soil nutrients or water, extreme soil acidity or alkalinity, very high or low temperatures, air pollutants, or mechanical, pesticide, or other injury. Severity and type of injury vary with the stage of plant maturity during which the disturbance occurs and the plant part involved. Symptoms of some noninfectious diseases are often confused with those caused by fungi, viruses, bacteria, mycoplasmas, nematodes, and parasitic seed plants.

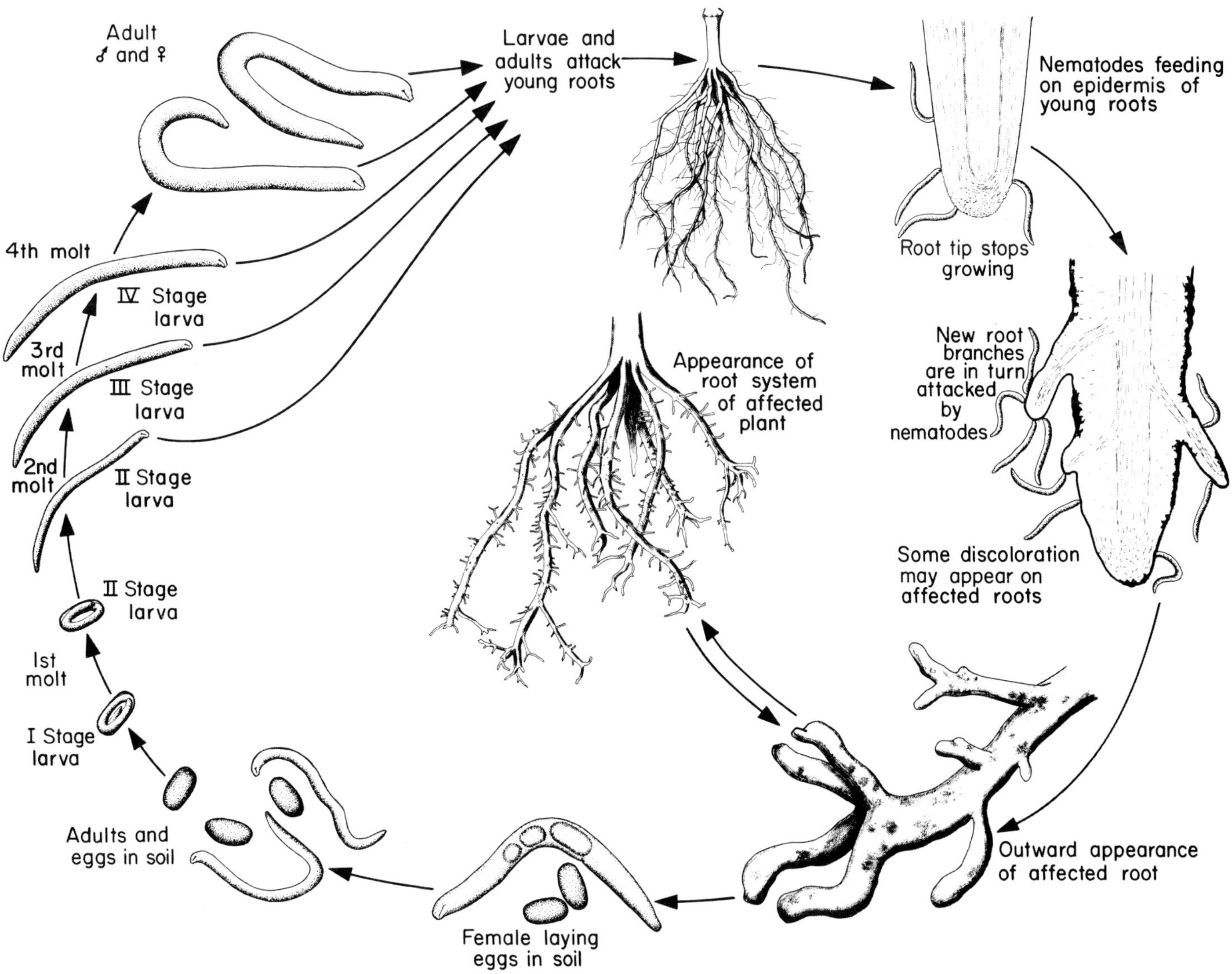

Fig. 5. Disease cycle of the stubby-root nematode *Trichodorus christii*. (Reprinted, by special permission, from Plant Pathology, 2nd ed., by G. N. Agrios. ©1978 The Academic Press, New York)

Host

Inbred lines, hybrids, and varieties differ greatly in disease resistance. In most cases, several genes determine resistance. They can be manipulated to produce varieties with high levels of disease resistance or tolerance and other desirable characteristics. In some cases, a single gene may determine resistance to a pathogen or a physiologic race of a pathogen.

Environment

Atmospheric and soil environments are important modifiers of disease development by affecting pathogen activity or host physiology. The most critical factors are air and soil temperatures, relative humidity, dew, precipitation, soil type, soil reaction (pH), and soil fertility. Cool, moist conditions favor development of certain diseases (e.g., yellow leaf blight and eyespot), while hot, dry weather favors others (e.g., common smut). Certain seed rots and seedling blights become problems at low soil temperatures; crazy top occurs in flooded or waterlogged soils.

Some stalk rots are particularly severe in infertile soils. Maintaining high, balanced soil fertility tends to produce vigorous plants that are more resistant to certain pathogens.

Some cultural practices (e.g., minimum or conservation tillage, sprinkler irrigation, and continuous cropping to maize) appear to increase disease severity by facilitating a buildup of inoculum in or near maize fields.

Maize Disease Control

The aim in control of maize diseases is to disrupt the combination of factors necessary for disease development. This is based on understanding the pathogens, disease cycles, plant parts involved, and the dissemination or spread of the causal organisms (pathogens).

Disease control may be achieved by a single procedure, but successful control of most diseases requires the application of several measures including the integrated use of environmental, genetic, and chemical factors.

Disease control procedures may be divided into four categories:

1. Disease Resistance: Planting adapted, disease-resistant hybrids and varieties is one of the most important and efficient means of control. Although no single variety or hybrid is resistant or tolerant to all diseases, some are resistant or highly tolerant to several major diseases in a specific area.
2. Chemical Control: Seed treatment with a fungicide, such as captan or thiram, controls some seed rots and seedling blights. Systemic fungicides are being used in certain areas of the world to control downy mildew diseases. Spraying the foliage with maneb, mancozeb (zinc ion coordinated maneb), zineb or other protective fungicide to control fungal leaf blights is usually economically feasible only in high-value, seed-producing fields and breeding nurseries.
3. Sanitation and Cultural Procedures: These include plowing under crop residues, rotating crops, maintaining well-balanced fertility based on soil test, preparing proper seedbeds, sowing disease-free seed, and controlling weeds and insects. The incidence of Goss' bacterial wilt and blight, Cercospora or gray leaf spot, anthracnose, eyespot, yellow leaf blight, and southern corn or maydis leaf blight, for example, can be partially reduced by destroying diseased crop residues.

   Nutrient balance is the key to the fertility effect in the development of some diseases, e.g., certain stalk rots and leaf blights are often most severe where there is too much nitrogen (N) and too little potassium (K).
4. Exclusion of the Pathogen: This can be achieved through quarantines and embargoes to keep out pathogens not yet in a given area or country.

Control measures of a general nature are listed for each disease or group of diseases in this Compendium.

Part I. Infectious Diseases of Maize

Bacterial Diseases

Bacteria comprise a group of some 4,000 procaryotic species, most of which lack chlorophyll and are saprophytic. Several hundred species cause diseases of plants or animals or both. Bacteria are found in air, water, and soil and also on or in all plants, animals, and man. They are remarkably variable in their characteristics, widely adaptable, and commonly persist in heterogenous populations. Their classification and identification can be difficult. Maize pathogens are identified from symptoms, cell and colony morphology, host specificity, and reaction to a wide variety of serological and biochemical tests.

Plant-pathogenic bacteria are normally unicellular, nonspore-forming rods up to 3 μm in length. Many species have one to many flagella that aid in motility via rotary or gliding motion. The most common types of bacteria divide by binary fission; populations of the resulting cells may separate or form associations of cells such as angular arrangements. In a warm, moist environment, large numbers of new cells can be produced within a few hours.

Bacteria are disseminated by man through a wide range of cultural practices and by transporting diseased plant material. Animals (including insects, mites, nematodes), splashing or flowing water, and windblown sand or soil are other common disseminating agents. Water-soaked tissues often predispose plants to invasion by bacteria. Free moisture and moderate-to-warm temperatures are general requirements for pathogen and disease development.

Bacteria enter plants through *wounds* produced by adverse weather, insects, nematodes, or man's activities, or through *natural openings* such as stomates, lenticels, hydathodes, nectaries, and leaf scars.

Bacteria can multiply rapidly inside plants where they may cause death of cells (necrosis), abnormal growth (e.g., tumors), blockage of water-conducting tissue (wilting), or breakdown of tissue structure (e.g., soft rots). Mycoplasmalike organisms (page 11) commonly occur in the food-conducting tissue. Bacteria often migrate throughout the plant, i.e., become systemic. Bacteria are pathogenic primarily through the action of enzymes or toxins that result in chlorosis, water-soaking, and other symptoms or signs.

When conditions are unfavorable for growth and multiplication, bacteria remain dormant on or within living or dead plants, soil, tools and equipment, or the bodies of insects.

Most plant-pathogenic species are quickly killed by high temperatures (10 min at 51°C when exposed), dry conditions, and sunlight. Few pathogenic bacteria survive in soil. Some are ingested by protozoa and other minute animals, or may be inhibited by antibiotics secreted by soil-inhabiting microorganisms including actinomycetes, other bacteria, and fungi.

Selected References

BRADBURY, J. F. 1970. Isolation and preliminary study of bacteria from plants. Rev. Plant Pathol. 49:213-218.

BUCHANAN, R. E., and N. E. GIBBONS. 1974. Bergey's Manual of Determinative Bacteriology (8th ed.). Williams & Wilkins: Baltimore, MD.

ELLIOTT, C. 1951. Manual of Bacterial Plant Pathogens (2nd ed.). Chronica Botanica: Waltham, MA.

SCHUSTER, M. L., and D. P. COYNE. 1974. Survival mechanisms of phytopathogenic bacteria. Annu. Rev. Phytopathol. 12:199-221.

STARR, M. P. 1959. Bacteria as plant pathogens. Annu. Rev. Microbiol. 13:211-238.

Stewart's Bacterial Wilt

Stewart's bacterial wilt, sometimes called Stewart's leaf blight or maize bacteriosis, is common in eastern North America and has been reported in Mexico, Costa Rica, Puerto Rico, eastern Europe, Italy, USSR, and China.

Symptoms

Infected sweet maize hybrids wilt rapidly, resembling plants suffering from drought, nutritional deficiency, or insect injury. Leaves show linear, pale green to yellow streaks with irregular or wavy margins that parallel veins and may extend the length of the leaf (Fig. 6). These streaks soon become dry and brown. Infected plants may produce premature, bleached, and dead tassels. Cavities may form in the stalk pit of severely infected plants near the soil line (Fig. 7). In such plants, bacteria spread throughout the vascular system, sometimes passing into the kernels.

Although most field or dent maize is not as susceptible as sweet maize, there are some very susceptible inbred lines and hybrids. Seedlings of very susceptible dent hybrids and lines may wither and die. Symptoms are short to long, irregular, pale green to yellow streaks in the leaves. This is known as the leaf-blight or late-infection stage because it generally occurs after tasseling. The streaked areas, which die and become straw-colored (Fig. 8, Plate C-1), originate from feeding marks of the corn flea beetle. Sometimes entire leaves die and dry up. When leaves die prematurely, yield is reduced and the weakened plants become more susceptible to stalk rots.

Causal Organism

Erwinia stewartii (Smith) Dye 1963

The bacterium is a nonmotile, nonflagellate, Gram-negative rod, 0.4–0.8 × 0.9–2.2 μm. Colonies on nutrient-glucose agar are cream-yellow, lemon-yellow, or orange-yellow and flat, raised, or convex, respectively. Colonies may be "running" or fluidal.

Masses of bacteria may ooze as yellow, moist beads from cut ends of infected stalks or may "stream" from the cut edges of infected leaf tissue.

Disease Cycle

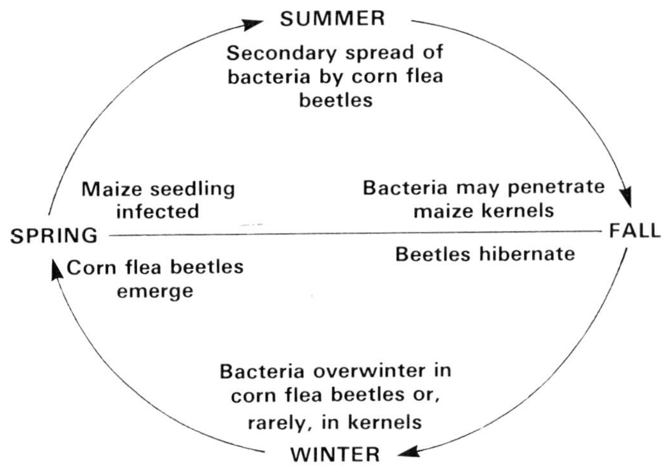

Host-Parasite Relations

In sweet maize, bacteria may be found in roots, stalks, leaf blades and sheaths, tassels, cobs, husks, and kernels. Dent maize kernel infection is rare and occurs only where the disease is severe on very susceptible inbred lines.

Insect Vectors

The corn flea beetle (*Chaetocnema pulicaria* Melsh.) (Fig. 9), both adult and larva of the twelve-spotted cucumber beetle (*Diabrotica undecimpunctata howarti* Barb.), toothed flea beetle (*C. denticulata* Ill.), as well as larvae of the seed corn maggot (*Hylemya cilicrura* Rond.), wheat wireworm (*Agriotes mancus* Say), and May beetle (*Phyllophaga* sp.) are known vectors of *E. stewartii.* Only *C. pulicaria* is known to be important in overwintering and spread of the bacteria.

Epidemiology

Mineral nutrition influences the susceptibility of maize hybrids to infection by *E. stewartii.* High levels of ammonium N and phosphorus (P) increase susceptibility, whereas high calcium (Ca) and K levels tend to decrease susceptibility. Disease severity is aggravated by high temperatures. Teosinte and eastern gamagrass are also

Fig. 6. Stewart's bacterial wilt of sweet maize. Streaks with wavy or irregular margins turn from pale green to yellow to brown. (Courtesy A. J. Ullstrup)

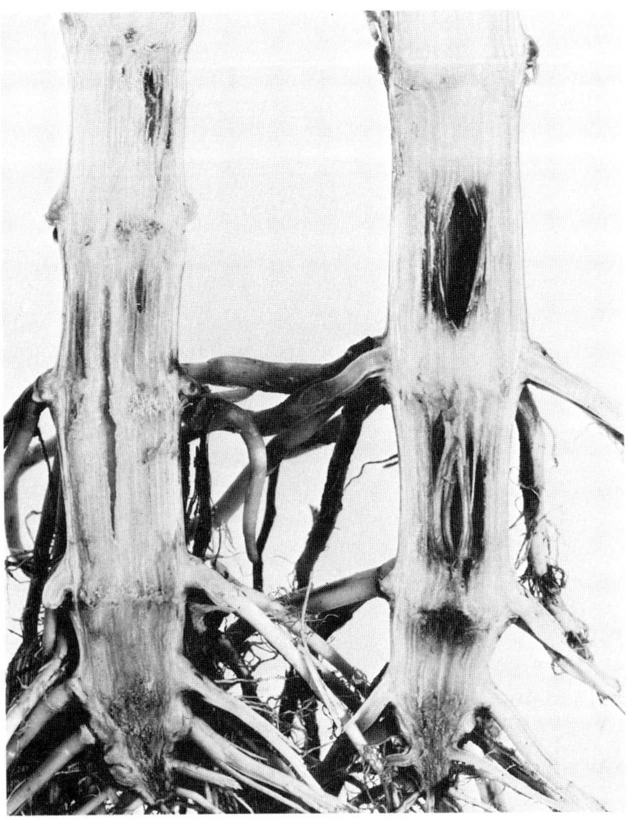

Fig. 7. Chocolate brown cavities may form in the stalk pith of plants severely infected with Stewart's bacterial wilt.

susceptible to *E. stewartii*. Dent maize generally resists infection better than sweet maize. Root and stalk rots are associated with severe infections of sweet and dent maize.

Disease Forecasting

Disease forecasting is based on the sum of the mean temperatures (°C) for December, January, and February. In mild winters (sum of mean temperatures above 37–38°C), large numbers of beetles survive, start infections, and spread bacteria during the growing season. Cold winters (sum of mean temperatures below 32°C) reduce beetle populations and limit disease development and spread.

Fig. 8. Bacterial leaf blight (late Stewart's bacterial wilt) generally appears on field or dent maize after tasseling. Note irregular or wavy margins of the lesions.

Fig. 9. Adult corn flea beetle, vector of Stewart's bacterial wilt bacterium. (Courtesy A. J. Ullstrup)

Control

1. Resistant hybrids and varieties.
2. Early applications of insecticides (e.g., carbaryl) to kill corn flea beetles.

Selected References

BUCHANAN, R. E., and N. E. GIBBONS. 1974. Bergey's Manual of Determinative Bacteriology (8th ed.). William & Wilkins: Baltimore, MD.

PEPPER, E. H. 1967. Stewart's Bacterial Wilt of Corn. Monograph No. 4. Am. Phytopathol. Soc.: St. Paul, MN.

ROBERT, A. L. 1967. Bacterial wilt and Stewart's leaf blight of corn. U.S. Dep. Agric. Farmers Bull. 2092.

Goss' Bacterial Wilt and Blight

This disease, sometimes called leaf freckles and wilt, has been observed on dent maize in central and eastern Nebraska since 1969. It has spread north and west in Nebraska, and was found in western Iowa in 1971 and 1978, eastern Iowa in 1979, several Kansas counties since 1973, and in South Dakota and Colorado since 1974. It has not been reported outside the United States. Losses in yield up to 50% have been reported.

Symptoms

The gross morphological symptoms (wilt and leaf blight) are indistinguishable from Stewart's bacterial wilt (see above). Discrete lesions, containing water-soaked streaks, appear parallel to leaf veins. As the streaks enlarge, droplets of bacterial exudate may appear on the surface. The droplets soon dry, leaving a crystalline sheen. Lesions may occur on seedlings or older plants and may arise from direct leaf infection or via roots and stalks of young plants. Early infection may cause seedlings to wilt, wither, and die. Later infection results in stunting, wilting, or various degrees of leaf blight that appear as gray to light greenish-yellow stripes, occasionally reddish on certain hybrids or inbreds, with wavy or irregular margins that follow the leaf veins and with or without dark green to black, water-soaked, angular spots (freckles) along the leaf veins. Eventually, the entire leaf dries up. Systemically infected plants may have discolored vascular bundles. A dry or water-soaked to slimy brown rot of the roots and lower stalk may occur, depending on the moisture regime. Plants can be infected, wilt, and die at any stage of development. An orange bacterial exudate is discharged from the vascular bundles when a stalk is cross-sectioned (Plate 2).

Causal Organism

Corynebacterium nebraskense (Schuster, Hoff, Mandel, and Lazar 1973) emend. Vidaver and Mandel 1974.

The bacterium is a nonmotile, nonflagellate, Gram-positive, pleomorphic rod, averaging 0.5–2.5 μm. The color of the colony varies with the medium. On nutrient broth-glucose-yeast extract agar, colonies are apricot or capucine orange and mucoid. Numerous strains of the bacterium are known. A medium selective for *C. nebraskense* is available.

Disease Cycle

The pathogen overwinters in infected maize leaves, stalks, cobs, and ears on or near the soil surface. An additional source of primary inoculum is infected kernels. The bacterium also survives readily in irrigation water. Injury is required for transmission of the pathogen. Infection may be direct in leaves or via the roots and stems. Wounding from sand-blasting, hail, severe rainstorms, and wind favors infection.

Epidemiology

Many of the details are not known. The pathogen infects dent, sweet, flint, and popcorn maize varieties and has been isolated from natural infections of green foxtail and shattercane. Seedlings are more susceptible than older plants. Additional plants found susceptible following artificial inoculation include grain sorghum, eastern gamagrass, Sundangrass, sugarcane, and teosinte. Symptoms are a red vein discoloration and long red or small irregularly shaped lesions following the leaf veins.

Control

1. Resistant dent and sweet maize hybrids are available.
2. Crop rotation.
3. Selection of seed from noninfested areas.
4. Deep-plowing of infected maize debris, preferably after harvest.

Selected References

CALUB, A. G., W. A. COMPTON, C. O. GARDNER, and M. L. SCHUSTER. 1974. Reaction of 113 corn (*Zea mays*) genotypes to leaf freckles and wilt. Plant Dis. Rep. 58:956-960.

GROSS, D. C., and A. K. VIDAVER. 1979. A selective medium for isolation of *Corynebacterium nebraskense* from soil and plant parts. Phytopathology 69:82-87.

SCHUSTER, M. L. 1975. Leaf freckles and wilt of corn incited by *Corynebacterium nebraskense* Schuster, Hoff, Mandel, Lazar, 1972. Nebr. Agric. Exp. Stn. Res. Bull. 270.

SCHUSTER, M. L., B. HOFF, and W. A. COMPTON. 1975. Variation, maintenance, loss, and failure to recover virulence in *Corynebacterium nebraskense*. Plant Dis. Rep. 59:101-105.

VIDAVER, A. K., and M. MANDEL. 1974. *Corynebacterium nebraskense*, a new, orange-pigmented phytopathogenic species. Int. J. Syst. Bacteriol. 24:482-485.

Holcus Spot

Symptoms

Round to elliptical spots, ranging from 2 to 10 mm in diameter, appear toward the tips of lower leaves. At first the spots are dark green and water-soaked; later they become creamy-white to tan and eventually dry and turn brown, often with reddish to brown margins. Larger lesions may be surrounded by a yellowish halo (Fig. 10).

Causal Organism

Pseudomonas syringae van Hall 1902 (Syn. *P. holci* (Kendrick) 1926)

The pathogen is a short rod with rounded ends, 0.6–1.2 × 1.5–3 or more μm (av. 0.73 × 2.13 μm), Gram-negative, and motile with one to several polar flagella, sometimes in long chains. Causal bacteria stream out of infected leaf tissue when mounted in water and viewed through a microscope. Isolation is easily accomplished on King's B medium agar by plating surface-sterilized leaf tissue first ground in sterile water or saline. Direct puncture with a sterile dissecting needle onto agar is also successful if developing or young lesions are sampled. Colonies are round, smooth, glistening, raised or pulvinate, grayish-white by reflected light, and greenish-fluorescent by transmitted light.

Epidemiology

Bacteria overwinter in crop residue and invade the host through stomata. *P. syringae* attacks leaves of several grasses including foxtail millet, pearl millet, Sudangrass, broomcorn, Johnsongrass, and some sorghum varieties. Other crops are infected by different strains of *P. syringae,* some of which appear to be host specific but of little economic importance. Warm (25–30° C), rainy, and windy weather, especially early in the season, favors disease development.

Control

1. Resistant hybrids and varieties.
2. Crop rotation.

Selected Reference

KENDRICK, J. B. 1926. Holcus bacterial spot of *Zea mays* and *Holcus* species. Iowa Agric. Exp. Stn. Bull. 100:303-334.

Fig. 10. Holcus spot caused by *Pseudomonas syringae.* Round to elliptical spots are dark green and water-soaked at first, later become dry and creamy white to tan. (Courtesy H. G. Johnson)

Bacterial Stripe and Leaf Spot

Symptoms

Typical primary symptoms are amber-colored to olive-colored, oil-soaked, translucent lesions with parallel sides that tend to elongate and coalesce.

Symptoms appear first on the lower leaves and, under favorable conditions, spread upward. Leaves above the ear are seldom infected. Lesions later become brown and necrotic. Lesions may appear as spots with irregular margins. A few very susceptible inbred lines show chlorotic striping or bleaching of the upper leaves (Fig. 11). Occasionally, the disease is concentrated in certain areas of a field. The disease is widespread but of little economic importance.

Causal Organism

Pseudomonas andropogonis (Smith) Stapp 1928

The bacterium is a short rod with rounded ends, 0.5–0.8 × 1.5–2.5 μm, Gram-negative, nonacid-fast, and motile with a polar flagellum. Causal bacteria stream out of infected tissue when mounted in water and observed with a microscope. No fluorescent pigment is produced on King's B medium agar.

Epidemiology

Like other bacterial diseases, bacterial stripe and leaf spot becomes more severe under extended periods of warm, wet weather. Bacteria enter the host through stomata when leaves are water-soaked. *P. andropogonis* also invades sorghum species, including Sudangrass and Johnsongrass.

Fig. 11. Chlorotic striping and bleaching caused by bacterial stripe. (Courtesy A. J. Ullstrup)

Control

Resistant hybrids and varieties.

Selected References

ULLSTRUP, A. J. 1960. Bacterial stripe of corn. Phytopathology 50:906-910.

VIDAVER, A. K., and R. R. CARLSON. 1978. Leaf spot of field corn caused by *Pseudomonas andropogonis*. Plant Dis. Rep. 62:213-216.

Bacterial Leaf Blight

Bacterial leaf blight has been reported throughout the central and southern Corn Belts of the United States.

Symptoms

Water-soaked linear lesions occur on the leaves as they emerge from the whorl. Lesions turn brown and then gray or white (Plate 3). Lesions seldom elongate after the leaf matures. New lesions are rarely observed after the plant tassels. Badly infected leaves often shred easily after they mature.

Causal Organism

Pseudomonas avenae Manns 1905 (Syn. *P. alboprecipitans* Rosen 1922)

The pathogen is a Gram-negative, polar-flagellated rod approximately 0.6 × 1.6 μm. Colonies on yeast extract-dextrose-calcium carbonate agar are cream with tan-to-brown centers, convex, and smooth. They reach a diameter of 2–3 mm after three days' incubation at 30° C. Optimum temperature for growth is 36° C. No fluorescent pigment is produced on King's B medium agar.

Epidemiology

The bacterium attacks several other plants, including oats, foxtails, and Vasey grass. Vasey grass in Florida (USA) serves as a primary source of inoculum. Warm, rainy weather favors development of the disease but it will occur at temperatures below 18–20° C. The bacterium does not survive well in soil or plant debris.

Control

Resistant hybrids and varieties.

Selected References

JOHNSON, A. G., A. L. ROBERT, and L. CASH. 1949. Bacterial leaf blight and stalk rot of corn. J. Agric. Res. 78:719-732.

SCHAAD, N. W., and B. M. CUNFER. 1979. Synonymy of *Pseudomonas coronafaciens, Pseudomonas coronafaciens* pathovar *zeae. Pseudomonas coronafeciens* subsp. *atropurpurea* and *Pseudomonas striafaciens.* Int. J. Syst. Bacteriol. 29:213-221.

SCHAAD, N. W., C. I. KADO, and D. R. SUMNER. 1975. Synonymy of *Pseudomonas avenae* Manns 1905 and *Pseudomonas alboprecipitans* Rosen 1922. Int. J. Syst. Bacteriol. 25:133-137.

SUMNER, D. R., and N. W. SCHAAD. 1977. Epidemiology and control of bacterial leaf blight of corn. Phytopathology 67:1113-1118.

Chocolate Spot

This leaf blight disease occurs sporadically in Wisconsin (USA). It has not been reported elsewhere. The bacterium also induces halo blight in oats and timothy (*Phleum pratense*).

Symptoms

Dark brown elongated spots up to 3 cm long, surrounded by a broad yellow halo, develop on the leaves (Plate 4). Lesions are most numerous along the leaf edges and toward the tip areas where K deficiency symptoms (page 82) also first appear. Lesions may later coalesce causing death of large areas.

Causal Organism

Pseudomonas atrofaciens pathovar *zeae*

The bacterium is a Gram-negative rod, approximately 0.6 × 1.0–2.7 μm that may form short chains, produces fluorescent pigment in King's B medium containing glycerol, is motile by one to five polar flagella, and has an oxidative mode of glucose metabolism. It forms levan, is isoxidase-negative and arginine dihydrolase-negative, and induces a hypersensitive reaction when infiltrated into tobacco leaves. The organism differs from the Holcus spot bacterium because it cannot utilize DL-lactate nor induce necrotic lesions on cowpea, and differs by its tyrosinase activity.

Epidemiology

Chocolate spot only occurs in K-deficient soils. Wounding of leaves by wind-whipping appears to promote infection.

Control

Apply a K-containing fertilizer based on a soil test.

Selected Reference

RIBEIRO, R. L. D., R. D. DURBIN, D. C. ARNY, and T. F. UCHYTIL. 1977. Characterization of the bacterium inciting chocolate spot of corn. Phytopathology 67:1427-1431.

Other Bacterial Leaf Spots and Blights

Several bacteria occur on maize leaves causing water-soaked, elliptical, or irregular lesions that may coalesce to form large necrotic areas. Stunting and shredding of leaves may occur, especially after wind and rain.

Bacterial Stalk Rot

Symptoms

Primary symptoms generally appear in midseason when plants suddenly fall over. One to several internodes above the soil line appear tan to dark brown, water-soaked, soft or slimy, and collapsed. Infected tissue produces a foul odor. Affected plants may remain green for several days because the vascular strands remain intact. A top and stalk rot occurs in maize plants sprinkler-irrigated with river, lake, or impounded water. The tips of the uppermost leaves wilt, followed by a slimy soft rot at the base of the whorl. The decay spreads rapidly downward until the plants collapse. Collapsed, twisted stalks are a good indication of this disease (see also Pythium stalk rot, page 46). Bacterial stalk rot is uncommon and limited in extent (Fig. 12 and Table III, page 44).

Causal Organism

Erwinia chrysanthemi pathovar *zeae* (Syn. *E. carotovora* var. *zeae* Sabet 1954)

The bacterium is a short, Gram-negative rod, 0.6–0.9 × 0.8 – 1.7 μm, actively motile with peritrichous flagella, usually in pairs and rarely in short chains, noncapsulated, and nonacid-fast. On nutrient agar, the colonies are grayish-white, raised, glistening, and smooth with entire margins. On potato-glucose agar (pH 6.5), the colonies are characteristically umbonate with undulate to coralloid margins ("fried egg") at 3–6 days' growth.

Disease Cycle

The bacterium lives saprophytically on crop residue in the soil and invades maize plants through stomata,

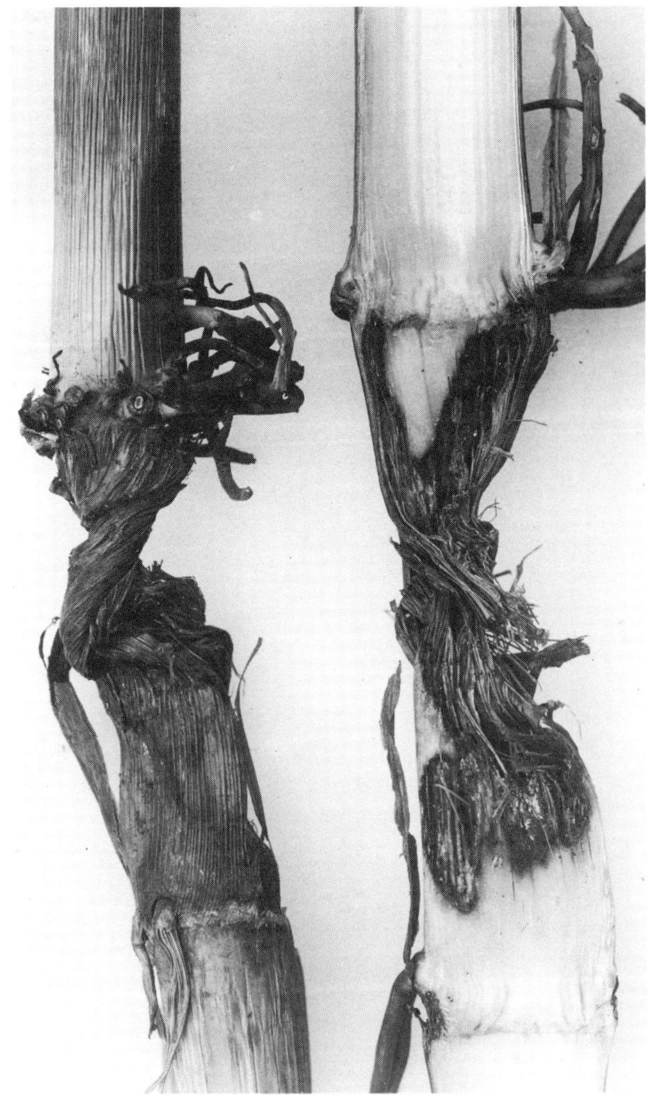

Fig. 12. Bacterial stalk rot. The stalk on the right has been split to show the internal decay. Affected stalks suddenly collapse and are usually twisted.

hydathodes, or wounds in the leaves or stalks. The organism is also seedborne.

Epidemiology

Bacterial stalk rot is most prevalent and destructive in areas with high rainfall, where plants are watered by sprinkler irrigation, and on land subject to flooding. The disease is favored by high temperatures (30–35° C) and poor air circulation.

Control

1. Resistant hybrids and varieties.
2. Good cultural management to avoid flooding.

Selected Reference

CHRISTENSEN, J. J., and R. D. WILCOXSON. 1966. Stalk Rot of Corn. Monograph No. 3. Am. Phytopathol. Soc.: St. Paul, MN.

Mycoplasma Diseases

Mycoplasmas and spiroplasmas are minute, filterable bacteria in the class Mollicutes. They are procaryotic, one-fifth to one-half the size of commonly known bacteria, and lack a rigid cell wall. Mycoplasma cells are pleomorphic. They are typically round bodies 300–800 nm in diameter to slender branched filaments of uniform diameter (100–800 nm), ranging in length from a few to 150 μm. Mycoplasmas are the smallest, free-living organisms capable of multiplication by binary fission and fragmentation of filaments on artificial media in the laboratory.

Spiroplasmas, like mycoplasmas, lack a cell wall and are bounded by a trilaminar membrane. Spiroplasmas, however, characteristically grow as mobile, helical filamentous cells.

Evidence from electron microscopy and antibiotic treatments indicates that organisms closely resembling "classical" mycoplasmas are involved in many plant diseases previously thought to be caused by viruses. Most of these organisms have not yet been cultivated in vitro, are incompletely characterized, and have not been shown conclusively that they are mycoplasmas. Until more is known, they are generally referred to as mycoplasmalike organisms (MLO). Most diseases now thought to be caused by MLO or spiroplasmas are the so-called yellows-type diseases. Characteristic symptoms include yellowing, shortening of the internodes, phyllody, proliferation of axillary shoots to produce a witches'-broom growth, sterility, virescence, and reduction of root growth.

Like some viruses, mycoplasmas and spiroplasmas occur only in phloem cells of a diseased plant and are transmitted by insects (leafhoppers), dodder (*Cuscuta* spp.), grafting, and propagation of infected plant parts. Unlike viruses (pages 60–69), however, they are sensitive to the tetracycline class of antibiotics, and antibiotic treatment of diseased plants causes temporary remission of symptoms.

Cell wall-less procaryotes (mycoplasmas and spiroplasmas) have thus far been found associated with only three (two?) diseases of maize, Rio Grande maize stunt, Mesa Central maize stunt, and maize bushy stunt. These diseases, and others now known to be induced by certain viruses (e.g., maize chlorotic mottle virus, page 63), were formerly collectively referred to as corn stunt. Development of resistant hybrids and varieties appears to offer the best hope of control for these diseases.

Corn Stunt or Maize Stunt

This major disease limits production of maize throughout the tropical lowlands of Latin America, being most severe in Mexico and other countries in Central America. It is found as far south as the lowlands of Bolivia. In newly infested areas, yield losses of 25–100% may occur. Corn stunt was first observed in the early 1940s in the Rio Grande Valley of Texas. The disease is known as maize stunt (achaparramiento) of the Rio Grande type. The disease was widespread throughout the Valley and was more common in sweet and popcorn than in field corn. Texas is the only state in the USA where the presence of corn stunt has been confirmed.

Because the principal leafhopper vector (*Dalbulus maidis* DeLong & Wolcott) (Fig. 13) occurs in Alabama, Arizona, California, Florida, Georgia, Louisiana, Mississippi, and New Mexico, the possibility exists that maize stunt may be present in these states.

Symptoms

The first symptoms are chlorosis of the margins of the whorl leaf followed by reddening of the tops of older leaves (Plate 5). These symptoms are followed by small chlorotic spots that develop at the base of newly developing leaves. In successive leaves above those bearing the first symptoms, the spots coalesce to form chlorotic stripes that extend toward the leaf tips until entire leaves are affected. Diseased plants are stunted and bear numerous small ear shoots (Plate 6).

Causal Organism

A helical, motile, cell wall-free procaryote not yet assigned a Latin binomial but clearly a member of the genus *Spiroplasma*. Cells are about 0.15–0.2 μm in diameter, 2 to about 15 μm in length, and 0.4–0.6 μm in

Fig. 13. This leafhopper (*Dalbulus maidis*) transmits the agents causing corn or maize stunt, maize bushy stunt, and maize rayado fino. (Courtesy L. R. Nault)

amplitude of the helical gyre, with a regular gyre length in a given helical filament. Cell dimensions vary slightly with medium in vitro cultivation. Helical cells exhibit flexional and rotational motility, with translational movement in viscous media. In culture, acid is produced from glucose and fructose; sterols are necessary for growth. Optimum temperature for growth is near 30°C. Growth is inhibited by homologous antiserum and by tetracycline antibiotics but not by penicillin.

Host-Parasite Relations

In infected plants, the spiroplasma cells are found only in phloem tissue. The spiroplasma is distributed throughout the diseased plant and occurs in leaves, stems, roots, ears, and tassels. In infected leafhopper vectors, the spiroplasma is found in various organs. Although helical cells are found in the hemolymph, morphology of the spiroplasma in other tissues may be largely nonhelical. The lifespan and fecundity are reduced by the spiroplasma infection in some species of leafhopper vectors (e.g., *Dalbulus elimatus* Ball), and cytopathology is observed in brain and salivary glands. The minimum incubation period of the corn stunt spiroplasma (CSS) in its leafhopper vectors is 14 days.

Disease Cycle and Epidemiology

Maize and its wild teosinte relatives (*Zea perennis* and *Z. mays mexicana*) are the only known hosts for CSS. Although *Z. perennis* and *Z. mays mexicana* are known only as experimental hosts, it seems plausible that at least *Z. mays mexicana* is a host of CSS in nature. In regions where corn is grown year-round, corn leafhoppers move from maturing diseased fields to newly planted fields, carrying the CSS with them. How the leafhopper vector and the CSS span corn-cropping seasons in regions where corn is not grown continuously is not known. In nature, vectors of CSS include the leafhopper species *D. maidis* and *D. elimatus* (Fig. 13). Several other leafhopper species are less efficient vectors; these include *Baldulus tripsaci* Kramer & Whitcomb and *Graminella nigrifrons* (Forbes), which occurs abundantly in the Corn Belt and the southern states of the USA.

Control

Development of resistant hybrids and varieties appears to offer the best hope for control. Resistance is polygenic and additive in nature.

Maize Bushy Stunt

This disease was first described from stunted maize plants collected from Texas in 1975. The disease is thought by some researchers to be the same as or closely related to Mesa Central corn (maize) stunt from Mexico and Central and South America and the corn stunts reported from Louisiana and Mississippi (USA). The disease may also occur in other southern US states where its efficent leafhopper vectors, *Dalbulus maidis* and *D. elimatus,* are found.

Symptoms

The first symptom is chlorosis of the margins of the whorl leaves followed by intense reddening of the tips of older leaves, although some maize varieties do not redden (Plate 7). All leaves that appear later develop chlorosis of the margins, yellowing or reddening, tearing, twisting, and are shortened. Numerous tillers may develop at the leaf axils and base of an affected plant giving it a bushy appearance. Infected plants are stunted and develop several small ears (Plate 8).

Causal Organism

A mycoplasmalike organism (MLO) has been found associated with maize bushy stunt (and Mesa Central stunt) disease. The organism appears to be distinct from the helical cell wall-free organism (spiroplasma) that causes corn stunt (Rio Grande type maize stunt). The maize bushy stunt MLO has not been cultured in vitro. Tetracycline treatment of inoculated plants causes a delay in disease development and prevents leafhoppers from transmitting the MLO.

Host-Parasite Relations

The MLO is found only in the phloem of infected plants. Symptoms may appear as early as eight or nine days after seedlings are inoculated when plants are held at 31°C. Symptom development may take two or three weeks at lower temperatures or when older plants are inoculated. The MLO is not mechanically transmissible.

Disease Cycle

Several leafhopper species transmit the MLO but only *D. maidis* and *D. elimatus* are efficient vectors. Maize and annual teosinte are the only known hosts for the maize bushy stunt causal agent. How or where the MLO or its leafhopper vectors survive in regions where maize is not grown continuously is not known.

Control

No control measures have been developed.

Selected References

CHEN, T. A., and C. H. LIAO. 1975. Corn stunt spiroplasma: Isolation, cultivation, and proof of pathogenicity. Science 188:1015-1017.

DAVIS, R. E. 1976. Spiroplasma: Role in the diagnosis of corn stunt disease, pp. 92-96. Proc. Int. Maize Virus Dis. Colloq. Workshop, Wooster, OH.

DAVIS, R. E., and R. F. WHITCOMB. 1971. Mycoplasmas, rickettsiae, and chlamydiae: Possible relation to yellows diseases and other disorders of plants and insects. Annu. Rev. Phytopathol. 9:119-154.

GRANADOS, R. R., and D. J. MEEHAN. 1975. Pathogenicity of the corn stunt agent to an insect vector, *Dalbulus maidis*. J. Invertebr. Pathol. 26:313-320.

NAULT, L. R., and O. E. BRADFUTE. 1979. Corn stunt: Involvement of a complex of leafhopper-borne pathogens, pp. 561-586. In: MARAMOROSCH, K., and HARRIS, K. (eds.). Leafhopper Vectors and Plant Disease Agents. Academic Press: New York.

TULLY, J. G., and S. RAZIN. 1977. The Mollicutes, pp. 405-459. In: LASKIN, A. I. and LECHEVALIER, H. A. (eds.). Handbook of Microbiology. Vol. 1, 2nd ed. CRC Press: Cleveland, OH.

WILLIAMSON, D. L., and R. F. WHITCOMB. 1975. Plant mycoplasmas: A cultivable spiroplasma causes corn stunt disease. Science 188:1018-1020.

Fungal Diseases

Fungi comprise a diverse group of lower plants that lack chlorophyll, usually have chitinous cell walls (cellulose in the oomycetes), and grow as saprophytes or parasites. They are broadly adapted for survival in air, soil, and water. More than 100,000 species of fungi have been described, probably less than half the total number in the world. Approximately 20,000 species are pathogenic to plants or animals or both. Most plant-pathogenic fungi are facultative saprophytes capable of growth on a variety of culture media and on nonliving plant tissues. Other fungi are highly specialized obligate parasites that grow and reproduce only in intimate association with a narrow range of living plant hosts. These include the rusts, many smuts, and the powdery, downy, and black mildews.

Most fungi produce branched filaments (hyphae) up to 100 μm wide. The hyphae grow and branch to form a fungus body (mycelium). The mycelium may be an interlacing tangle of hyphae, a loose woolly mass, or a compact body.

Nearly all species reproduce, spread, and survive by means of asexual or sexual spores. The spores are easily disseminated by air currents, splashing or flowing water, on or in plant parts or animals, and by the activities of man. Resting spores, e.g., chlamydospores and oospores, allow certain fungi to survive unfavorable growing conditions. Some fungi do not produce spores. They multiply and overseason by forming compact masses of hyphae (sclerotia), or the fungus body fragments and these are spread by water, wind, man, and other agents. Spore morphology and the mechanisms of spore production are basic to fungal taxonomy and nomenclature.

Fungal hyphae invade a maize plant through *wounds* made by man, hail, wind, blowing sand, insects, nematodes, or other fungi; through *natural openings*; or by *direct penetration* using pressure or enzyme action or both.

Fungi cause the majority of infectious diseases of maize including the rusts; smuts; downy mildews; most rots, spots, and blights; and deformations. Symptoms of disease result from the depletion of host nutrients, the action of toxic or growth-stimulating by-products, and the mechanical displacement of host tissues. Many fungal diseases are diagnosed by structures produced by the pathogens.

Fungi, like bacteria, overseason on and in living or dead plants, soil, or occasionally in insects.

Species of fungi, like other types of pathogens, are subdivided into strains or physiologic races that may differ in such characteristics as the species, varieties, and cultivars of plants they attack; the degree of virulence; environmental parameters at which infection and disease development occur; and the ability to develop in lower levels of soil.

Selected References

AINSWORTH, G. C., and A. S. SUSSMAN. 1965. The Fungi—An Advanced Treatise. 4 vols. Academic Press: New York.

ALEXOPOULOS, C. J. 1979. Introductory Mycology (3rd ed.). Wiley: New York.

BARNETT, H. L., and B. B. HUNTER. 1972. Illustrated Genera of Imperfect Fungi (3rd ed.). Burgess: Minneapolis, MN.

Seed Rots and Seedling Blights

Germinating maize kernels may be attacked by a number of soilborne or seedborne fungi that cause seed rots and seedling blights. Severe infection may kill the embryo before germination (seed rot) or destroy the seedling before or after emergence. The terms "preemergence and postemergence damping-off" are often used to specify the affected growth stage. These diseases are prevalent in poorly drained, excessively compacted, or cold (less than 10–13°C) and wet soils. Disease severity is affected by planting depth, soil type, age, and quality of the seed, mechanical injury to the pericarp, and genetic resistance to infection. Sweet maize is more susceptible than dent maize.

Symptoms

Above-ground symptoms of these diseases are yellowing, wilting, and death of the leaves (Plate 9). They may be confused with mechanical or chemical injury or insect damage. Examination of parts in the ground is therefore necessary for accurate diagnosis.

Seed Rot. Rotting of seed before germination.

Damping-Off and Seedling Blight. Soft rot of stem tissues (mesocotyl) near the ground level and water-soaking of seedling tissues. The rotted area may be dark with sporangia and oospores in the tissues (*Pythium*), or with whitish-gray (*Diplodia*), white-to-pink (*Fusarium*), or bluish (*Penicillium*) mycelium and masses of spores (Fig. 14).

Seedling Wilt. A gray coloration starts at the leaf tips and extends rapidly to the whole leaf, causing complete collapse of seedlings in 24-48 hr. This symptom is suggestive of kernels infected with *Helminthosporium*

Fig. 14. Penicillium seedling blight. Dead areas are blue-green at first, then turn drab or light brown; healthy leaf at left.

maydis (Race T), especially when no fungus can be isolated from infected leaves. Wilting may also occur following infection by *Penicillium oxalicum* Currie & Thom.

Root Rots

No sharp demarcation between seedling blight and root rot exists, and many of the same causal organisms attack both stem and root tissues. Root rots of maize are widely distributed but apparently do not cause loss except in special situations.

Late-season root rots, such as those associated with *Fusarium* species, increase with maturity, senescence of root tissues, and field stresses. They are commonly a prelude to stalk rot (page 43).

Early-season root rots occur in soils where oxygen is deficient due to poor drainage or compaction. *Pythium graminicola* Subram. is often the primary cause up to the mid-whorl growth stage and again at the dent stage, along with other fungi. *Helminthosporium pedicellatum* Henry (*Exserohilum pedicellatum*) (Henry) Leonard & Suggs can be isolated throughout the season and *Fusarium roseum* f. sp. *cerealis* 'Graminearum' and *Pyrenochaeta terrestris* (Hansen) Gorenz, Walker & Larson commonly late in the season.

Symptoms

Small, yellowish-brown lesions occur on the primary roots and later on the first and second whorls of secondary roots. The affected roots later become black and necrotic. Above-ground symptoms such as wilting, stunting, and yellowing of leaves may accompany root rot but these symptoms are not always apparent.

Disease Cycle

The disease cycle of Pythium root rot and seedling disease is illustrated in Fig. 15. Sporangia and oospores are produced within or outside of host tissue. Germ tubes or mycelia contact the seeds, seedling tissues, or root tips of maize either by chance or by chemical attraction. The mycelium proliferates in young tissue, causing cell collapse and death. On entering the roots, the fungus progresses toward the stem or seed, growing rapidly between and through the cells and killing the roots. The mesocotyl region is very susceptible, and rotting in this area severs the plumule from the seed. Severe postemergence seedling losses can occur when the top

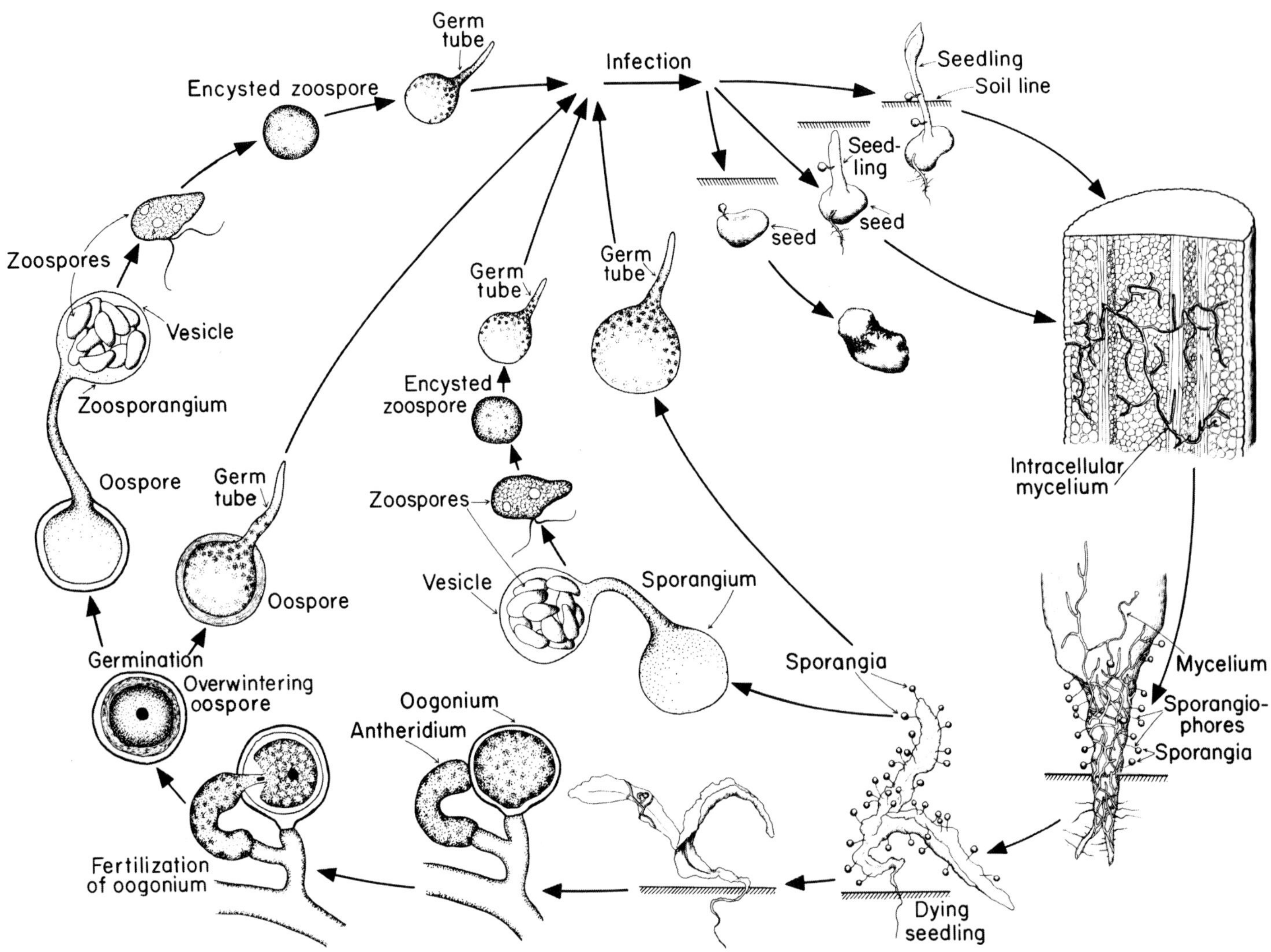

Fig. 15. Pythium disease cycle. (Reprinted, by special permission, from Plant Pathology, 2nd ed., by G. N. Agrios.©1978 The Academic Press, New York)

2–4 cm of soil are dry and the adventitious roots fail to develop.

Causal Organisms of Seed Rots, Seedling Blights, Seedling Wilt, and Root Rots

The most important organisms are species of *Pythium,* including *P. irregulare* Buis., *P. debaryanum* Hesse, *P. graminicola* Subram., *P. splendens* Braun, *P. vexans* dBy., *P. rostratum* Butl., *P. paroecandrum* Drechs., and *P. ultimum* Trow. These soilborne fungi produce white, slender mycelia with asexual sporangia of various shapes. *Pythium* species are found throughout the world in water, muck, or heavy soils in association with dead organic matter or in the roots of susceptible plants. Sporangia germinate directly by germ tube formation or by means of a vesicle containing biflagellate zoospores. Sexual reproduction results in thick-walled oospores that serve as the overwintering or oversummering stage of the fungus. Other fungi attacking the seed, seedlings, and roots include: *Fusarium moniliforme* Sheld., *Diplodia maydis* (Berk.) Sacc., *F. roseum* Lk. ex Fr. f. sp. *cerealis* 'Graminearum,' *Helminthosporium maydis* Nisik., *H. pedicellatum* Henry (*Exserohilum pedicellatum* (Henry) Leonard & Suggs), *Phymatotrichum omnivorum* (Shear) Duggar, and *Colletotrichum graminicola* (Ces.) G. W. Wils.

Also associated with seedling blights are *Aspergillus* spp.; *Nigrospora oryzae* (Berk. & Br.) Petch; *Rhizoctonia zeae* Voorhees; *Macrophomina phaseolina* (Tassi) G. Goid., and *Cephalosporium maydis* Samra, Sabet & Hingorani. Many of these fungi also cause ear and kernel rots and are discussed more fully in that section (page 51).

Control

1. Injury-free seed of high germination.
2. Good cultural practices, e.g., planting seed in warm, fairly moist soil (above 12.8°C); proper seedbed preparation; correct placement of fertilizer, herbicide, and other pesticides.
3. Seed-protectant fungicides, e.g., captan or thiram.

Selected References

FUTRELL, M. C., and M. KILGORE. 1969. Poor stands of corn and reproduction of root growth caused by *Fusarium moniliforme.* Plant Dis. Rep. 53:213-215.

KOEHLER, B. 1957. Pericarp injuries in seed corn. Ill. Agric. Exp. Stn. Bull. 617.

KOMMEDAHL, T., and D. S. LANG. 1971. Seedling wilt of corn from kernels naturally infected with *Helminthosporium maydis* in Minnesota. Plant Dis. Rep. 55:371-373.

MOHAMED, H. A., W. E. ASHOUR, A. R. SIRRY, and S. M. FATHI. 1968. Fungi causing seedling blight of corn in the United Arab Republic. Plant Dis. Rep. 52:84-86.

SHEPHERD, R. J., E. E. BUTLER, and D. H. HALL. 1967. Occurrence of a root rot disease of corn caused by *Helminthosporium pedicellatum.* Phytopathology 57:52-56.

Helminthosporium Leaf Spots and Blights

TAXONOMIC NOTE: Mycologists now generally agree that the genus name *Helminthosporium* is inappropriate when applied to any of the species of maize pathogens traditionally included in this genus, e.g., *Helminthosporium maydis, H. carbonum, H. turcicum, H. rostratum,* and *H. pedicellatum.* However, they are not in agreement as to what the disposition of all these species should be, the alternatives being the genera *Bipolaris* Shoemaker, *Drechslera* Subramanian & Jain, and *Exserohilum* Leonard & Suggs. Because the preponderance of world literature, both classic and modern, is found under the genus name *Helminthosporium,* we have chosen to retain this name in this edition of the Compendium.

Southern or Maydis Leaf Blight

Symptoms

Lesions on the leaves caused by Race O are elongated between the veins, tan, 2–6 × 3–22 mm long, with limited parallel margins and buff-to-brown borders (Plate 10). Variation in lesion size and shape may occur between inbreds and hybrids with different genetic backgrounds. Race O usually attacks only the leaves. Lesions produced by Race T are tan, 0.6–1.2 × 0.6–2.7 cm, spindle-shaped or elliptical, with yellow-green or chlorotic halos (Fig. 16 and Plate 11). Later, the Race T lesions often have dark, reddish-brown borders and may occur on the leaves, stalks, leaf sheaths (Plate 12), ear husks, shanks, ears, and cobs. A black, felty mold may cover the affected kernels. With Race T, ear and cob rot can occur, with substantial losses in harvesting and shelling. Early shank invasion can cause premature death of the ear and possible ear drop.

Seedlings from infected kernels (Race T) may wilt and die within three to four weeks after planting.

Early death of leaves caused by either race predisposes the plants to stalk rot.

The two races are morphologically similar but Race T is specifically pathogenic to maize containing Texas male-sterile cytoplasm (Tm-sc) while Race O is not. Race T produces a pathotoxin that is highly toxic only to susceptible maize (Tm-sc), whereas Race O produces only a small amount of toxin nonspecific to cytoplasms.

Cultural isolates of Race O, but not Race T, sporulate well on V-8® juice agar. When the substrate is supplemented with L-asparagine, Race T sporulates nearly as well as Race O.

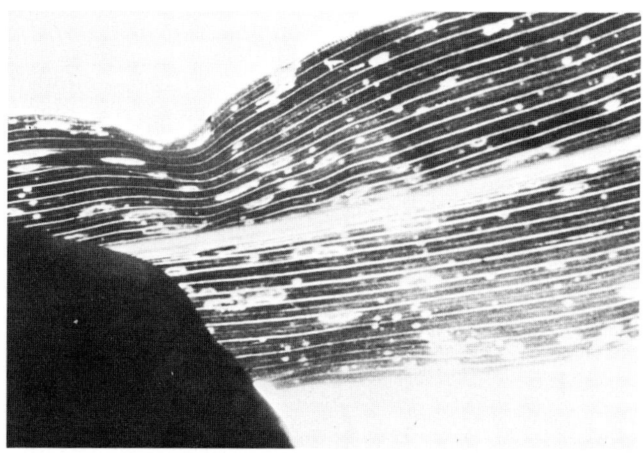

Fig. 16. Close-up of race T lesions of southern leaf blight on a leaf with Texas male-sterile cytoplasm. Characteristic yellow halos appear when held to the light.

Causal Organism

Helminthosporium maydis Nisik. (Syn. *Bipolaris maydis* (Nisik.) Shoemaker, *Drechslera maydis* (Nisik.) Subram. & Jain.) Perfect state is *Cochliobolus heterostrophus* (Drechs.) Drechs.

The conidiophores are long, 120–170 μm, curved, and grow out through the stomata. The conidia are olive-green, 10 – 17×30 – 115 μm, curved, tapering toward the ends, 3 to 13 septate, and germinate by polar germ tubes. The hilum does not protrude (Fig. 17B). Spores can be obtained by placing infected tissue in a moist chamber for 24–48 hr, taking care not to saturate the tissue.

Pseudothecia of *C. heterostrophus,* which occur rarely in nature, appear as tiny specks on or in maize tissue. The asci in the pseudothecia typically contain four (1–8) filamentous, five to nine-septate ascospores, 6–7 × 130–340 μm arranged in parallel coils.

Disease Cycle

The fungus overwinters as mycelium and spores in maize debris in the field and Race T on kernels in cribs, bins, and elevators. The conidia are carried by wind or splashing water to growing plants, where primary infections occur. Sporulation on the lesions produces additional primary or secondary inoculum. The disease cycle can be completed in about 60–72 hr (Race T) under ideal conditions. Kernels infected with Race T may be a potential means of overwintering and spread of *H. maydis.* Infected grain is *not* toxic to livestock and other animals.

Epidemiology

Southern corn leaf blight occurs worldwide, being important in regions of warm (20–32° C), damp climate. Race T develops in more northern areas than the older Race O. Long dry periods of sunny weather between rains is unfavorable for disease development.

Races O and T of *H. maydis* are difficult to identify by leaf symptoms alone on normal cytoplasm maize. Race T is virulent on maize with Tm-sc, attacking husks and ears as well as leaves and inciting small lesions on normal maize. Race O shows no differential reaction between plants with different cytoplasms. Male-sterile cytoplasms other than Tm-sc are resistant to Race T. Race O resistance is controlled by a single recessive gene, *rhm,* but resistance may be polygenically determined.

Control

1. Hybrids with normal cytoplasm are resistant to Race T. Genetic sources resistant to both Races O and T are available. Tm-sc hybrids differ in their susceptibility to Race T. Maize with Tm-sc is no longer in commercial use.

2. Clean plowdown of crop debris may reduce early infections.

3. Where practical in seed-production fields, fungicides may be applied, starting when one or two lesions occur per leaf and before the disease becomes epidemic.

Northern or Turcicum Leaf Blight

Symptoms

Long, elliptical, grayish-green or tan lesions ranging from 2.5 to 15 cm in length develop first on the lower leaves (Plate 13). The lesions vary in their morphology when gene HT_2 is utilized. The HT_2 gene conditions a long, chlorotic, linear streak that may extend the length of the leaf on some host genotypes; the lesion may be very easily mistaken with those caused by *Erwinia stewartii* (page 5). The disease progresses upward on the plant. Severe infection causes a prematurely dead and gray appearance that resembles frost or drought injury. In damp weather, large numbers of grayish-black spores are produced on the lesions, often in concentric or targetlike zones. The ears are not infected, although lesions may form on the outer husks.

Causal Organism

Helminthosporium turcicum Pass. (Syn. *Exserohilum turcicum* (Pass.) Leonard & Suggs, *Bipolaris turcica* (Pass.) Shoemaker, *Drechslera turcica* (Pass.) Subram. & Jain.) Perfect state is *Trichometasphaeria turcica* Luttrell. (Syn. *Setosphaeria turcica* (Luttrell) Leonard & Suggs.)

The spores are olive-gray, spindle-shaped, slightly curved, three to eight septate, 20 × 105 μm with a protruding hilum, and germinating by polar germ tubes. The conidiophores are olivaceous, two to four septate, and 7–9 × 150–250 μm (Fig. 17A). Sporulation can be induced in a moist chamber.

T. turcica, the sexual stage, occurs rarely, if ever, in nature but produces black, globose pseudothecia in the laboratory. The asci are cylindrical with a short stipe and contain one to eight, but usually two to four, ascospores, which are hyaline, straight or slightly curved, typically three septate, and 13–17 × 42–78 μm in size.

Two biotypes of *H. turcicum* from maize isolates have been described. One biotype is avirulent to maize carrying genes HT_1, HT_2, HT_3, and HtN. The other biotype is avirulent to maize carrying genes HT_2, HT_3, and HtN, but is virulent to maize carrying HT_1A or B.

Disease Cycle

H. turcicum overwinters as mycelia and conidia in infected leaves, husks, and other plant parts, except in the northernmost areas where maize can be grown. Conidial cells can be transformed into chlamydospores. Conidia are windborne over long distances to leaves of maize plants. Secondary spread within and between fields occurs by conidia produced abundantly on leaf lesions.

Epidemiology

Northern corn leaf blight occurs sporadically in most humid areas of the world where maize is grown. Disease development is favored by moderate temperatures (18–27° C) and heavy dews during the growing season. It is retarded by dry weather. If the disease is established before silking, losses in grain yield up to 50% may occur. If infection is moderate or delayed until six weeks after silking, yield losses are minimal.

H. turcicum attacks sorghum, Sudangrass, Johnson-grass, gamagrass, and teosinte. Races of *H. turcicum* on sorghum and Sudangrass apparently do not infect maize, but some maize isolates can infect Sudangrass.

At least two types of resistance are known: small lesion size and few lesions (polygenic), and chlorotic lesions with little or no sporulation and yellowish halos (monogenic). Monogenic resistance genes are HT_1, HT_2, and HT_3. Gene HtN conditions a delayed lesion expression in the host.

Control

1. Resistant hybrids and varieties.
2. Where practical in seed-production fields or on sweet corn for fresh market, fungicides may be applied, starting when lesions are first observed.

Helminthosporium Leaf Spot (Blight) and Northern Leaf Spot

Symptoms

Lesions of Race 1 are tan, oval to circular, usually with concentric zones and measure 1.2 × 2.5 cm. Race 1 of the causal fungus also readily attacks ears of susceptible

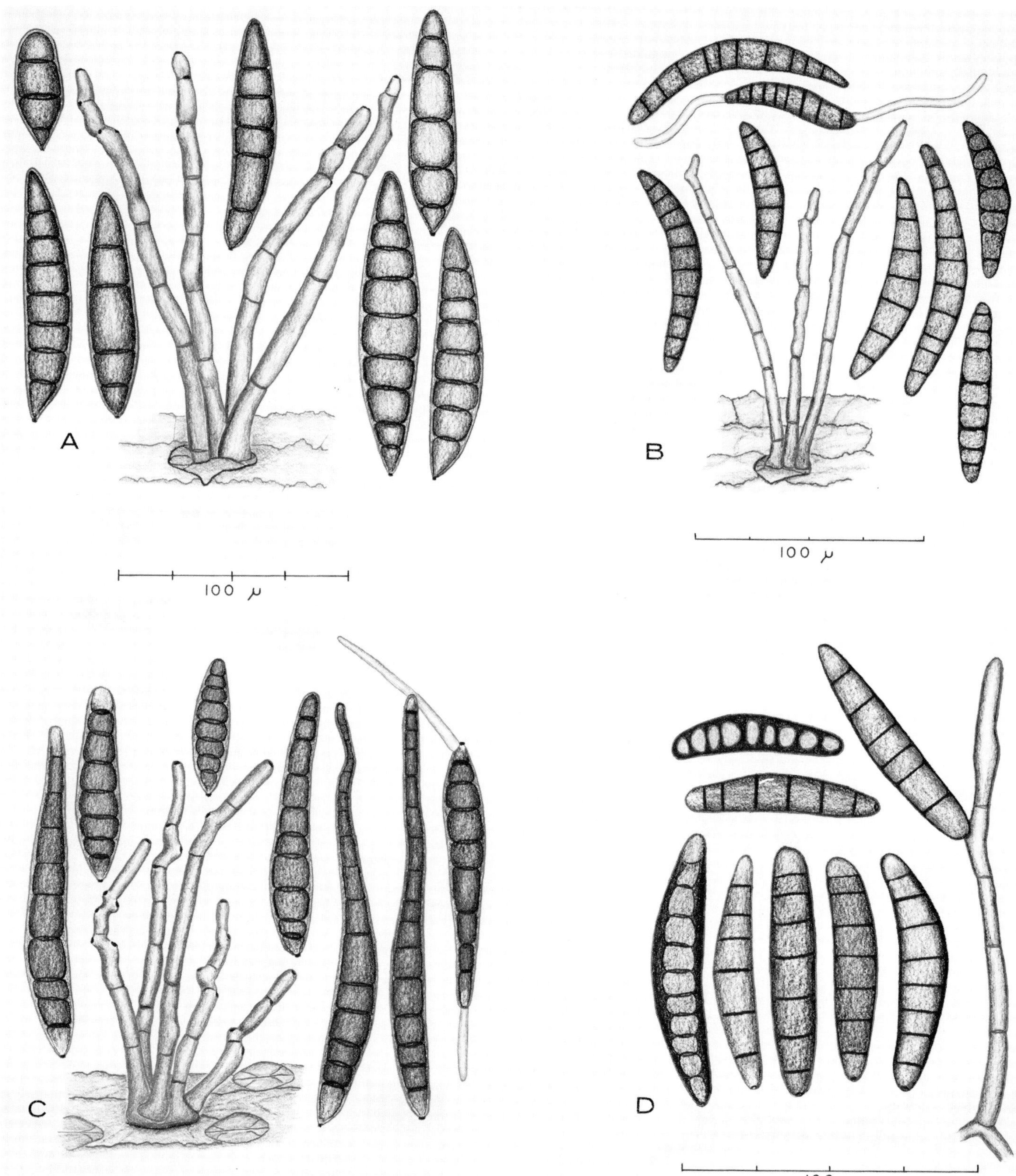

Fig. 17. Conidiophores and conidia of four species of *Helminthosporium:* **A,** *turcicum;* **B,** *maydis;* **C,** *rostratum;* **D,** *carbonum.* (A—C after Drechsler, D after Ellis)

plants causing a black, felty mold over the kernels. Race 2 produces oblong, chocolate-colored spots up to 0.5 × 2.5 cm, but this type of infection is rare. Ear infections caused by Race 2 are indistinguishable from those caused by Race 1 (Plate 14).

Race 3, called Helminthosporium leaf blight or Northern leaf spot, occurs widely in the northern Corn Belt of the USA. It produces narrow, linear lesions ranging in width from 0.5 to 2 mm and a maximum of 15 to 20 mm long on the leaf blades, sheaths, husks, and ears. The lesion type varies with the genotype of the host and the isolate. The lesions are grayish-tan and surrounded by a light to darkly pigmented border.

Causal Organism

Helminthosporium carbonum Ullstrup. (Syn. *Bipolaris zeicola* (Stout) Shoemaker, *Drechslera zeicola* (Stout) Subram. & Jain.) Perfect state is *Cochliobolus carbonum* Nelson.

The conidia are golden yellow to dark olive-brown, curved, long-elliptical to spindle-shaped, with rounded ends, 2 to 12 septate, 7–18 × 25–100 μm (Fig. 17D). Pseudothecia produced in pure culture are black, ellipsoid to globose, and contain cylindrical or club-shaped, straight or slightly curved asci. The ascospores are filiform, hyaline, with five to nine septa, 6.4–9.6 × 182–300 μm, and coiled in a close helix.

Disease Cycle

Similar to northern and southern leaf blights (see above).

Epidemiology

Moderate temperatures and high relative humidity favor the disease. Spores are produced abundantly in damp weather.

There are at least three races of *H. carbonum*. Race 1 is highly pathogenic on a few inbred lines; Race 2 is much less pathogenic with no distinct host specialization; and Race 3 occurs on commercial hybrids, but disease severity is minimal and of no apparent consequence in hybrid production. Race 3 has been a problem primarily in seed production with certain highly susceptible inbreds, and may be a biotype of Race 2.

Control

Resistant hybrids and varieties.

Helminthosporium Leaf Disease or Rostratum Spot

Symptoms

Lesions appear on mature plants as small, pale yellow, elongated spots 1–2 × 2–5 mm that gradually extend to form longitudinal bands or stripes between the veins. In severe cases, the lesions coalesce and extend across the veins. Mature lesions are straw-colored with light brown margins and measure 2–3 × 2–40 mm.

Causal Organism

Helminthosporium rostratum Drechs. (Syn. *Exserohilum rostratum* (Drechs.) Leonard & Suggs emended Leonard, *Bipolaris rostrata* (Drechs.) Shoemaker, *Drechslera rostrata* (Drechs.) Richardson & Fraser, *Exserohilum halodes* (Drechs.) Leonard & Suggs.) Perfect state is *Setosphaeria rostrata* Leonard.

The conidia are dark, olivaceous, straight or rarely curved, tapering toward both ends, with a prominent, dark, and protruding hilum, 3 to 15 septa, and 14–22 × 32–184 μm. The proximal and distal septa are often thicker and more prominent than the others (Fig. 17C).

Disease Cycle

Similar to northern and southern leaf blights (see above).

Epidemiology

Temperature and relative humidity are important factors in disease development. Infection is most severe at 30° C in a water-saturated atmosphere. Older tissue is more susceptible than young.

H. rostratum occurs also on the leaves of lovegrasses, sorghums, Sudangrass, pearl millet, and several other grass hosts.

Control

Resistant hybrids and varieties.

Selected References

BERGQUIST, R. R., and O. R. MASIAS. 1974. Physiologic specialization in *Trichometasphaeria turcica* f. sp. *zeae* and *T. turcica* f. sp. *sorghi* in Hawaii. Phytopathology 64:645-649.

BHOWMICK, T. P. 1966. *Helminthosporium rostratum* Drechs. on maize, Sudan grass and Johnson grass in India. Indian Phytopathol. Soc. Bull. 3:32-36.

DRECHSLER, C. 1923. Some graminicolous species of *Helminthosporium*. J. Agric. Res. 24:641-740.

DRECHSLER, C. 1925. Leafspot of maize caused by *Ophiobolus heterostrophus* n. sp., the ascigerous stage of a *Helminthosporium* exhibiting bipolar germination. J. Agric. Res. 31:701-726.

FISHER, D. E., A. L. HOOKER, S. M. LIM, and D. R. SMITH. 1976. Leaf infection and yield loss caused by four *Helminthosporium* leaf diseases of corn. Phytopathology 66:942-944.

HOOKER, A. L. 1972. Southern leaf blight of corn—present status and future prospects. J. Environ. Qual. 1:244-249.

HOOKER, A. L., A. MESTERHAZY, D. R. SMITH, and S. M. LIM. 1973. A Helminthosporium leaf blight of corn in the northern Corn Belt. Plant Dis. Rep. 57:195-198.

LEONARD, K. J. 1976. Synonymy of *Exserohilum halodes* with *E. rostratum*, and induction of the ascigerous state, *Setosphaeria rostrata*. Mycologia 67:402-411.

LEONARD, K. J., and E. G. SUGGS. 1974. *Setosphaeria prolata*, the ascigerous state of *Exserohilum prolatum*. Mycologia 68:281-297.

LEONARD, K. J., and E. G. SUGGS. 1974. Reaction of corn hybrids to *Exserohilum prolatum*. Plant Dis. Rep. 58:612-615.

LIM, S. M., J. G. KINSEY, and A. L. HOOKER. 1974. Inheritance of virulence in *Helminthosporium turcicum* to monogenic corn. Phytopathology 64:1150-1151.

NELSON, R. R., M. BLANCO, S. DALMACIO, and B. S. MOORE. 1973. A new race of *Helminthosporium carbonum* on corn. Plant Dis. Rep. 57:822-823.

ROBERT, A. L. 1962. New hosts for three *Helminthosporium* species from corn. Plant Dis. Rep. 46:321-324.

SMITH, D. R., A. L. HOOKER, and S. M. LIM. 1970. Physiologic races of *Helminthosporium maydis*. Plant Dis. Rep. 54:819-822.

SOUTHERN CORN LEAF BLIGHT SPECIAL ISSUE. December 1970. Part II. Plant Dis. Rep. 54:1099-1136.

ULLSTRUP, A. J. 1944. Further studies on a species of Helminthosporium parasitizing corn. Phytopathology 34:214-222.

ULLSTRUP, A. J. 1972. The impact of the southern corn leaf blight epidemics of 1970-1971. Annu. Rev. Phytopathol. 10:37-50.

WHITEHEAD, M. D., and O. H. CALVERT. 1959. *Helminthosporium rostratum* inciting ear rot of corn and leaf-spot of thirteen grass hosts. Phytopathology 49:817-820.

YOUNG, G. Y., C. L. LEFEBVRE, and A. G. JOHNSON. 1947. *Helminthosporium rostratum* on corn, sorghum, and pearl millet. Phytopathology 37:180-183.

Physoderma Brown Spot

Brown spot or Physoderma disease is generally of minor importance in the USA, although serious localized outbreaks have been reported. Severe infection occurs in Central America. The disease is primarily limited to areas of abundant rainfall and high mean temperatures.

Symptoms

Lesions first appear as very small, oblong to round, yellowish spots on the leaf blade (Plate 15), leaf sheath, stalk, and sometimes on the outer ear husks and tassels of maize. Lesions may occur in bands across the leaf blade. Infected tissues turn chocolate brown to reddish brown and coalesce to form large irregular, angular blotches (Fig. 18). Cells of infected tissues disintegrate and expose dusty pustules containing large numbers of golden brown to dark brown sporangia. Stalks infected at the nodes beneath the leaf sheaths frequently break at infection centers (Plate 16).

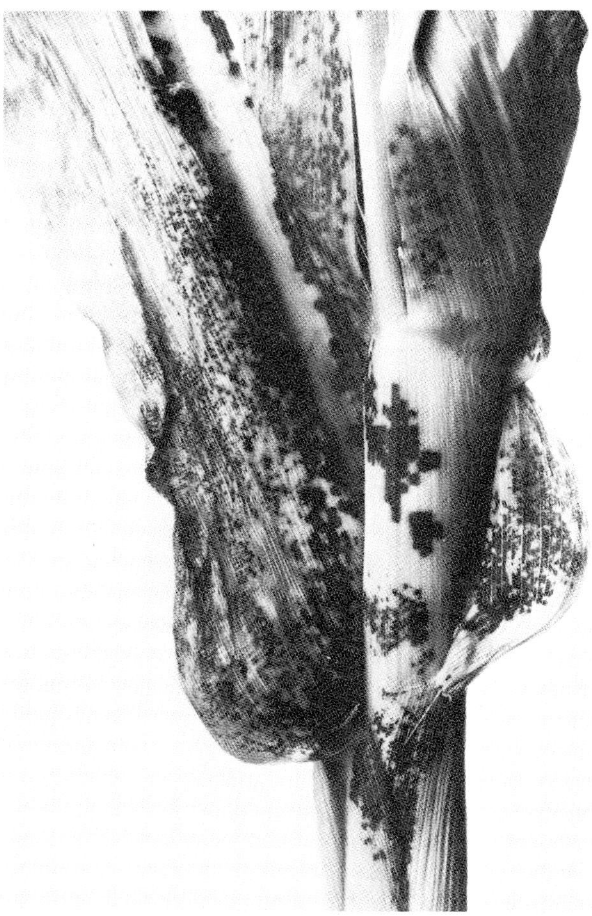

Fig. 18. Physoderma brown spot lesions start as small yellowish spots that later turn brown to reddish brown, merging to form large blotches.

Causal Organism

Physoderma maydis Miyabe (Syn. *P. zeae-maydis* Shaw)

Sporangia produced in pustules in infected tissue are smooth, brown, flattened on one side with a circular cap or lid, and 18–24 × 20–30 μm (Fig. 19B,C,D). On germination, which occurs only in light, a sporangium releases 20 to 50 zoospores, 3–4 × 5–7 μm with flagella three to four times the length of the spore. The zoospores germinate by producing fine fibrous hyphae. The mycelium is coenocytic.

Disease Cycle

The thick-walled sporangia overseason in infected tissue or in soil and germinate in the presence of moisture to produce numerous zoospores. Zoospores move in water on leaf surfaces, and attack young maize tissues usually within the leaf whorl, forming infectious hyphae. The resulting mycelium enters mesophyll or parenchyma cells forming larger vegetative structures from which sporangia develop 16 to 20 days after infection. Spread by secondary zoospores produced in thin-walled sporangia may be evident.

Epidemiology

Brown spot infection occurs generally in meristematic tissues in the whorl and is closely associated with the host's morphological development. Zoospores of *P. maydis* can infect host tissue only during certain periods of the day and within a few hours of their development. Infection occurs in a diurnal cycle resulting in alternating lateral bands of infected and noninfected leaf tissues. Sporangia require water on leaf surfaces and relatively high temperatures (23–30° C) for germination. Teosinte is also susceptible to *P. maydis*.

Control

1. Resistant hybrids and varieties.
2. Clean plowdown of crop debris.

Selected References

BROYLES, J. W. 1962. Penetration of meristematic tissues of corn by *Physoderma maydis*. Phytopathology 52:1013-1016.

HEBERT, T. T., and A. KELMAN. 1958. Factors influencing the germination of resting sporangia of *Physoderma maydis*. Phytopathology 48:102-106.

SPARROW, F. K. 1974. Observations on Chytridiaceous parasites of Phanerogams. XXI. The endobiotic stage of *Physoderma maydis*. Mycologia 66:693-697.

Yellow Leaf Blight or Phyllosticta Leaf Spot

Yellow leaf blight or Phyllosticta leaf spot appeared in the late 1960s primarily in the northern part of maize-growing areas of the USA and subsequently in several midwestern and southern states. It has also been identified in Argentina and Canada. Similar species have been reported from Europe, Africa, Brazil, and Taiwan.

Symptoms

Rectangular to oval, yellow, cream-colored or tan-colored lesions, sometimes surrounded by chlorotic tissue, first appear on the lower leaves especially on the outer half (Fig. 20 and Plate 17). Lesions vary in size, averaging 0.3 × 1.3 cm, and may coalesce causing leaves to be blighted. Leaf sheaths and outer husks are also susceptible. Pycnidia develop in the lesions and extrude conidia under moist conditions. Pycnidia, which appear as minute black specks, differentiate yellow leaf blight from most other foliage diseases. With heavy, early-season infections, leaves appear yellow, resembling those of N-deficient plants. Infected leaves turn brown and die. Plants may be stunted. Lodging caused by stalk-rotting fungi often follows severe yellow leaf blight.

Causal Organism

Phyllosticta maydis Arny & Nelson. Perfect state is *Mycosphaerella zeae-maydis* Mukunya & Boothroyd.

The fungus can be isolated and cultured on potato-dextrose agar (PDA) (22–25°C) after conventional surface sterilization of leaf tissue, and produces pycnidia under fluorescent light. Pycnidial formation can be induced by placing infected tissue in a moist chamber for 48–72 hr (Fig. 21). These pycnidia are subglobose, 60–150 μm in diameter, reddish brown, ostiolate, and immersed in necrotic tissue. The conidia

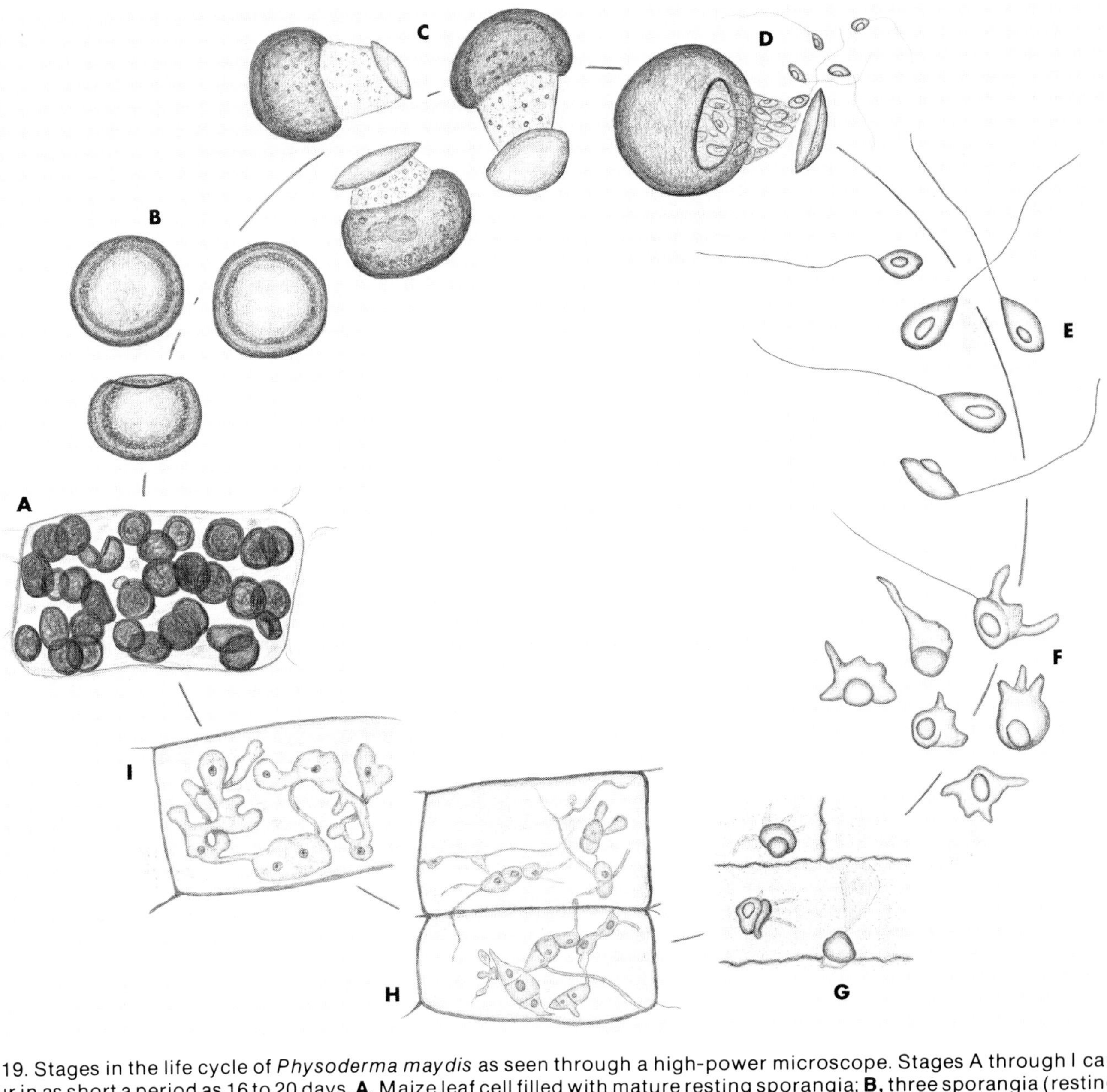

Fig. 19. Stages in the life cycle of *Physoderma maydis* as seen through a high-power microscope. Stages A through I can occur in as short a period as 16 to 20 days. **A,** Maize leaf cell filled with mature resting sporangia; **B,** three sporangia (resting spores), two top views and one side view; **C,** stages in opening of the sporangia showing early stages of zoospore formation. Note the dehisced operculum (lid) being carried up by the enlarging sporangium; **D,** mature zoospores escaping through the ruptured apex of the resting spore; **E,** zoospores with a single flagellum from a germinated resting spore; **F,** germinating zoospores, amoeboid stage; **G,** zoospores germinating by fine threadlike hyphae that have penetrated the epidermal cells of a maize leaf and have begun to form enlarged cells; **H,** Physoderma mycelium within two maize leaf cells showing the fibers and enlarged cells; **I,** rhizomycelium within a maize epidermal cell showing young sporangia at the ends of short hyphae. (A, D, E, F, G, and H after Tisdale; B and C after Sparrow; I after Wolf)

are single-celled, hyaline, oblong-ellipsoid to subcylindrical, and typically biguttulate, i.e., having two oil droplets. Conidial size is 3–7.5 × 8–20 μm, with the majority being 4–6 × 12–15 μm. (See Fig. 26E.)

Pseudothecia with mature ascospores form in culture at 21°C in darkness. The pseudothecia are about 140 μm in diameter, the asci 45–65 × 9.6–12 μm, and the ascospores 15–20 × 5–6 μm. The ascospores are hyaline and two-celled.

Disease Cycle

P. maydis overwinters in maize or grass debris. Susceptible host plants may be attacked at any growth stage. Immature pseudothecia are formed in debris in the fall and ascospores are produced in the spring. Conidia developing in pycnidia on large lesions provide secondary inoculum.

Epidemiology

Disease development is favored by cool, wet weather. Young plants growing through heavy surface debris of maize during such conditions are especially vulnerable. Reaction to yellow leaf blight is governed by genetic and cytoplasmic factors. As with Race T of Southern corn leaf blight, Texas male-sterile cytoplasm (Tm-sc) plants are more susceptible than plants with normal cytoplasm. Some other cytoplasms conferring male sterility do not increase susceptibility. Nuclear genes also affect the reaction. Inbreds are generally more susceptible than hybrids, making seed production fields more vulnerable to attack. Other hosts of *P. maydis* include foxtails and Sudangrass.

Control

1. Resistant hybrids and varieties offer the best means of control. Tm-sc cytoplasm is usually more susceptible than other cytoplasms.
2. Fungicide applications may be feasible in certain seed-production fields.
3. Clean plowing or rotation, or both, greatly reduces early-season infections and primary inoculum.

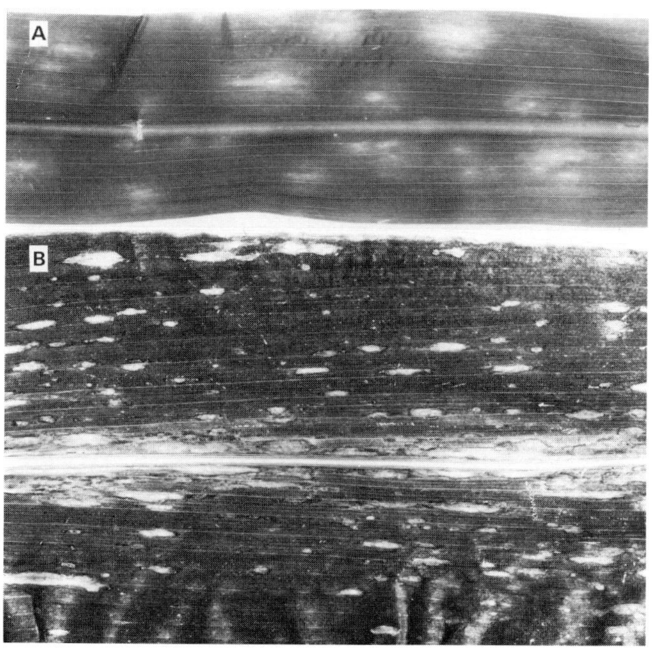

Fig. 20. Yellow leaf blight. **A,** Lesions on a young maize leaf with a wide chlorotic border; **B,** older lesions on a more mature leaf. (Courtesy G. L. Worf)

Selected References

ARNY, D. C., and R. R. NELSON. 1971. *Phyllosticta maydis* species nova, the incitant of the yellow leaf blight of maize. Phytopathology 61:1170-1172.

ARNY, D. C., G. L. WORF, R. W. AHRENS, and M. F. LINDSEY. 1970. Yellow leaf blight of maize in Wisconsin: Its history and the reactions of inbreds and crosses to the inciting fungus (*Phyllosticta* sp.). Plant Dis. Rep. 54:281-285.

CASTOR, L. L., J. E. AYERS, and R. R. NELSON. 1977. Controlled-environment studies of the epidemiology of yellow leaf blight of corn. Phytopathology 67:85-90.

JIMENEZ-DIAZ, R. M., and C. W. BOOTHROYD. 1976. Suscept-pathogen relationship in maize affected by yellow leaf blight. Phytopathology 66:1169-1173.

McFEELEY, J. C. 1971. Comparison of isolates causing yellow leaf blight of corn in Ohio. Plant Dis. Rep. 55:1064-1068.

MUKUNYA, D. M., and C. W. BOOTHROYD. 1973. *Mycosphaerella zeae-maydis* sp. n., the sexual stage of *Phyllosticta maydis*. Phytopathology 63:529-532.

NELSON, R. R., J. E. AYERS, and J. B. BECKETT. 1971. Reactions of various corn inbreds in normal and different male-sterile cytoplasms to the yellow leaf blight organism (*Phyllosticta* sp.). Plant Dis. Rep. 55:401-403.

SUTTON, J. C., A. BOOTSMA, and T. J. GILLESPIE. 1972. Influence of some cultural practices on yellow leaf blight of maize. Can. Plant Dis. Surv. 52:89-92.

Eyespot

Reported as "brown-spot" in Japan in 1959, eyespot has since been found in several north central states and Pennsylvania in the USA, Canada, Argentina, Austria, France, Germany, Yugoslavia, and New Zealand.

Symptoms

Early symptoms consist of small, translucent, circular to oval lesions (1–4 mm in diameter) with yellowish "halos" (Fig. 22A) that may fuse to form large necrotic areas. The spots usually develop early or late in the season in zones or patches. Initially the lesions are water-soaked and later develop tan-to-cream centers surrounded by a brown or purple ring with a narrow, yellow halo, giving the appearance of an "eyespot" (Fig. 22B and Plate 18). Lesions may develop on the leaf

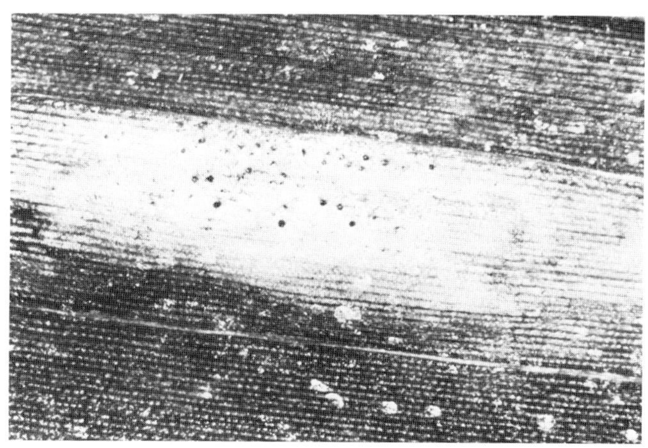

Fig. 21. Close-up of pycnidia of *Phyllosticta maydis* embedded in a yellow leaf blight lesion. (Courtesy G. L. Worf)

sheaths and outer husks, but are usually found in much larger numbers on the leaves of plants approaching maturity. Kernel infections have been reported on severely infected plants. The foliar symptoms may be confused with *Curvularia*, "physiological," or genetic leaf spots.

Causal Organism

Kabatiella zeae Narita & Y. Hiratsuka. Perfect state is *Aureobasidium zeae* (Narita & Y. Hiratsuka) Dingley.

The conidia are long, curved to falcate with pointed ends, hyaline and nonseptate, measuring 3–4 × 18–33 μm (av. 3.6 × 27 μm). Conidia are produced successively on short conidiophores, which develop through the stomata of maize leaves. Sporulation can be induced by placing diseased tissue in a humidity chamber for 24–48 hr. A more efficient method is to place a small rectangle of infected tissue upright in a petri dish of water agar with the lesion(s) above the agar and the dish placed in a moist chamber. Isolation can be made by streaking the spores or by placing young lesions on acidified PDA. On PDA at 24°C, growth is slow (2–3 mm per day), some colonies are yellow or pink turning dark blue to black with age, and growth is tough and leathery. The fungus also grows well on V-8 juice agar. A special liquid medium in shake culture for rapid spore production has been described.

Disease Cycle

The fungus overwinters in maize debris. Conidia produced on the debris are carried to nearby seedlings where they may germinate and cause infection. Lesions appear 4 to 10 days after leaf infection. Secondary spread is by rain splash and windborne spores. The fungus can also be seedborne.

Epidemiology

Only maize has been reported as a host for *A. zeae*. Disease development is favored by cool, humid weather.

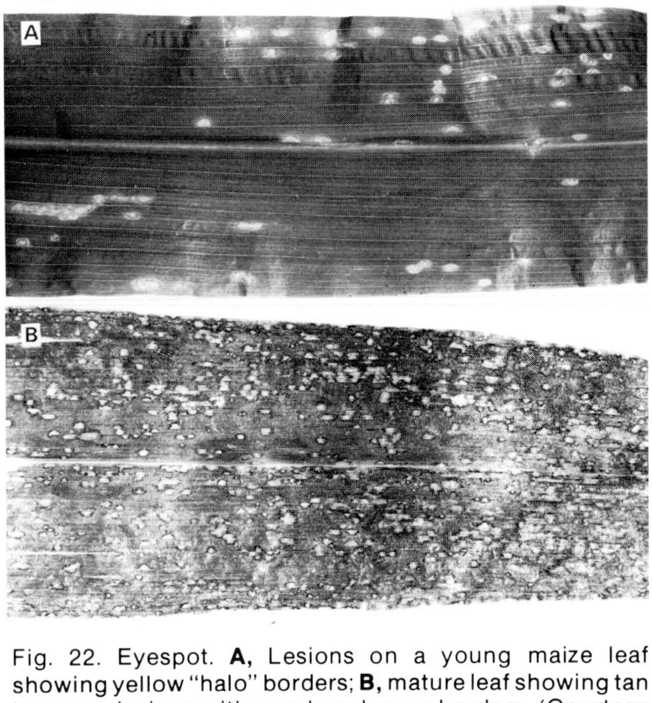

Fig. 22. Eyespot. **A,** Lesions on a young maize leaf showing yellow "halo" borders; **B,** mature leaf showing tan to cream lesions with purple or brown borders. (Courtesy G. L. Worf)

Certain inbred lines and hybrids appear to be less susceptible to the disease. Resistance is at least partially dominant. In culture, the optimum temperature for spore germination is 24°C.

Control

1. Less susceptible hybrids and varieties.
2. Clean plowing or crop rotation, or both, has reduced early disease infections.

Selected References

ARNY, D. C., E. B. SMALLEY, A. J. ULLSTRUP, G. L. WORF, and R. W. AHRENS. 1971. Eyespot of maize, a disease new to North America. Phytopathology 61:54-57.

DINGLEY, J. M. 1973. 'Eyespot' disease of maize in New Zealand. N. Z. J. Agric. Res. 16:325-328.

REIFSCHNEIDER, F. J. B., and D. C. ARNY. 1979. Seed infection of maize (*Zea mays*) by *Kabatiella zeae*. Plant Dis. Rep. 63:352-354.

REIFSCHNEIDER, F. J. B., and D. C. ARNY. 1979. A liquid medium for the production of *Kabatiella zeae* conidia. Can. J. Microbiol. 25:1100-1102.

SCHNEIDER, R., and W. KRÜGER. 1972. *Kabatiella zeae* Narita & Hiratsuka als erreger einer blattfleckenkrankeit an mais in Deutschland. Phytopath. Z. 74:238-248.

Anthracnose

Anthracnose has increased in importance in the USA since 1971. The disease is also important in France, Germany, The Philippines, Thailand, and other countries. Anthracnose leaf blight is common in fields early in the season where maize debris from the previous year has been left on the soil surface.

Symptoms

Symptoms vary greatly, depending on the genotype, age of the leaf, and environment. Small, oval to elongate, water-soaked spots first appear on the leaves at any stage of growth. The lesions are semitransparent and may originate on any part of the leaf blade. The spots may enlarge up to 15 mm long and become tan at the center with red, reddish-brown, or yellow-orange borders (Plate 19). The enlarging lesions may coalesce, blighting the entire leaf. Leaf symptoms are most common early in the season on the lower leaves and late

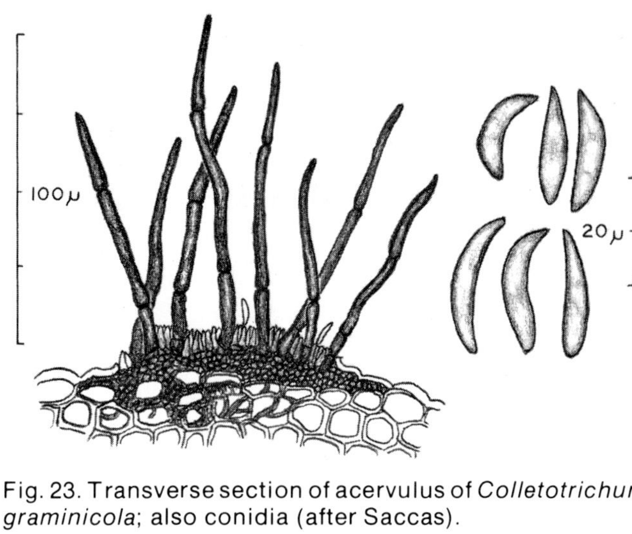

Fig. 23. Transverse section of acervulus of *Colletotrichum graminicola*; also conidia (after Saccas).

in the season on the upper leaves. Diseased leaves wither and die within a few days late in the season. Fruiting bodies (acervuli) may develop in abundance on dead tissues. Under suitable climatic conditions, these produce pinkish masses of conidia. Some cells of the fruiting body (acervulus) grow out into long, rigid, unbranched, dark, hairlike structures (setae) that attain lengths of 175 μm (Fig. 23) and rise above the surface of the leaf. The setae are easily observed with a hand lens. A preliminary field diagnosis for anthracnose is thus possible. On resistant plants, the lesions are usually smaller and are chlorotic to necrotic. On some genotypes, the lesions may be full-sized but may not expand to cause extensive leaf blighting. The fungus may also cause a severe top dieback and stalk rot (page 43).

Causal Organism

Colletotrichum graminicola (Ces.) G. W. Wils.

On PDA the mycelium is septate, sparingly branched, hyaline, and granular. Acervuli produced on agar and in dead host tissues are dark brown, circular to oval, thick, and, under moist conditions, appear as small specks scattered in infected tissues. Conidiophores produced in the acervuli are erect, hyaline, nonseptate, unbranched, short, and 1.6–3.3 × 4.9–13.3 μm. Long, dark-brown to black septate setae are produced among the conidiophores. The conidia, measuring 4.9–5.2 × 26.1–30.8 μm, are hyaline, nonseptate, cylindrical, become sickle-shaped with age, and are borne singly at the tips of the conidiophores (Fig. 23).

Disease Cycle

The fungus overseasons as a saprophyte on maize crop residue and seed as spores and mycelia. Spores from disintegrating debris are carried by wind or splashing rain to the leaves where penetration occurs directly through the epidermis or through stomata. Seedborne inoculum may cause seedling blight. Other grass hosts may serve as sources of primary inoculum. Spores produced on diseased plants provide secondary inoculum.

Host Range

The fungus affects a large number of grasses including the small grains, sorghums, bermudagrass, Dallis grass, Guineagrass, Bungalow grass, crabgrass, Guatemala grass, molasses grass, orchardgrass, Paragrass, redtop, red fescue, and Merker grass. Isolates (strains) from maize do not infect *Sorghum* spp. nor do isolates from *Sorghum* spp. infect maize.

Epidemiology

Anthracnose is generally favored by high temperatures and extended periods of cloudy weather. The disease develops best on older tissues, although seedlings may be infected. Free water is necessary for spore dispersion and germination. Strains of *C. graminicola* that differ in pathogenicity are known.

Control

1. Resistant hybrids and varieties. Resistance to leaf blight is not very highly correlated with resistance to anthracnose stalk rot (page 43).
2. Crop rotation and clean plowdown of crop residues.
3. Balanced soil fertility.

Selected References

DALE, J. L. 1963. Corn anthracnose. Plant Dis. Rep. 47:245-249.

MORGAN, O. D., and J. G. KANTZES. 1971. Observations of *Colletotrichum graminicola* on T corn and blends in Maryland. Plant Dis. Rep. 55:755.

POLITIS, D. J., and H. WHEELER. 1972. The perfect stage of *Colletotrichum graminicola*. Plant Dis. Rep. 56:1026-1027.

QUEBRAL, F. C. 1958. Anthracnose of corn. Philipp. Agric. 42:250-263.

WARREN, H. L., and R. L. NICHOLSON. 1975. Kernel infection, seedling blight and wilt of maize caused by *Colletotrichum graminicola*. Phytopathology 65:620-623.

WILLIAMS, L. E., and G. M. WILLIS. 1963. Disease of corn caused by *Colletotrichum graminicolum*. Phytopathology 53:364-365.

Other Fungal Leaf Diseases

Phaeosphaeria Leaf Spot

Symptoms

Lesions are initially small, pale green or chlorotic, becoming bleached or dried with dark brown margins. The spots are round, elongate to oblong, measuring 0.3–2.0 cm, and scattered over the leaf (Plate 20). Spots may coalesce and become irregularly shaped. Perithecia and, less frequently, pycnidia develop in the lesions.

Causal Organism

Phaeosphaeria maydis (P. Henn.) Rane, Payak & Renfro (Syn. *Sphaerulina maydis* P. Henn.)

The perithecia are spherical to subglobose with papillate ostioles. The asci are hyaline, clavate or cylindrical, straight or curved, eight-spored, and measure 44.5–70 × 7.5–8.5 μm. Ascospores are hyaline, straight or slightly curved, three septate, slightly constricted at the septa, and measure 14.5–17.5 × 3.5–5.0 μm. In culture, the fungus produces white mycelium that later becomes dark gray.

Pycnidial Stage

Pycnidia (*Phyllosticta* sp.) are spherical or globose, dark brown to black with a round ostiole. Spores are hyaline, ellipsoid, elongate or round, measuring 3.2–9 × 2.4–3.2 μm.

Disease Cycle

The fungus persists in diseased plant parts in the field. Under favorable conditions, spores germinate and infect maize leaves.

Epidemiology

The disease is favored by high rainfall and relatively low night temperatures. It is found in India, adjacent to the foothills of the eastern and central Himalayas, and in Costa Rica, Colombia, and Brazil.

Curvularia Leaf Spots

Lesions are at first small (1–2 mm), straw-colored, circular to oval, with reddish-brown or dark brown margins. They may coalesce to form necrotic areas up to 1 cm long. In the centers of the lesions caused by some species of *Curvularia,* circular to oval, grayish spots up to 1 mm in diameter appear. Straw-colored or chlorotic halos may surround the lesions (Fig. 24). Curvularia leaf spots are prevalent in hot, humid, maize-growing areas and can damage crops significantly.

Causal Organism

Curvularia spp.

Conidiophores are brown, septate, straight or flexuous, often bent, simple, or branched, and bear conidia sympodially.

C. lunata (Wakker) Boed. var. *aeria* (*Cochliobolus lunatus* Nelson & Haasis) is by far the most prevalent species. The conidia are similar to those of *C. pallescens,* measuring 18–32 × 8–16 μm (Fig. 25B).

C. pallescens Boed. The conidia are pale brown, usually straight or only slightly curved, with a distinct protuberant hilum, taper at both ends, usually three-septate, with the two middle cells being larger than the end cells, and measuring 17–32 × 7–12 μm (Fig. 25A).

Fig. 24. Curvularia leaf spot. (Courtesy B. L. Renfro)

Other species of *Curvularia* occurring on maize include:

1. *C. inaequalis* (Shear) Boed. Conidia measure 24–45 × 9–16 μm (Fig. 25C).
2. *C. maculans* (Bancroft) Boed.
3. *C. intermedia* Boed. (*Cochliobolus intermedius* Nelson). Conidia measure 27–40 × 13–20 μm (Fig. 25E).
4. *C. clavata* Jain. The three-septate conidia measure 18.5–35 × 7.5–12 μm (av. 24.9 × 8.4 μm).
5. *C. tuberculata* Jain. Conidia measure 23–52 × 13–22 μm (Fig. 25D).
6. *C. eragrostidis* (P. Henn.) J. A. Meyer. Conidia measure 22–33 × 10–18 μm (Fig. 25G).
7. *C. senegalensis* (Speg.) Subram. Conidia measure 19–30 × 10–14 μm (Fig. 25F).

Fusarium Leaf Spot

Field symptoms appear as water-soaked spotting mainly on the whorl leaves. Later, these spots appear white and papery with brown borders. The whorl does not open and leaf tips turn downward. Occasionally, necrotic spots develop on higher portions of the stalks. The disease is severe in dry, warm areas of islands in the Caribbean area. The causal fungus is *Fusarium moniliforme* (Fig. 51D), which also infects the stalks (page 45) and kernels (page 52) of maize.

Cercospora or Gray Leaf Spot

This disease occurs in warm-to-hot, humid areas of the USA, Mexico, Central America, northern South America, Europe, Africa, Southeast Asia, India, China, and The Philippines.

Symptoms

Lesions on maturing maize leaves are pale brown or gray to tan, long (0.5 to 5 cm), narrow and rectangular, being characteristically restricted by the veins. The lesions may coalesce, killing the leaves. The disease is usually first noticed attacking the lower leaves. Extensive leaf blighting may occur until all the leaves are killed, finally resulting in stalk breakage and lodging (Plate 21). Total loss may occur if gray leaf spot infection occurs early and favorable environmental conditions exist following infection.

Causal Organisms

Cercospora zeae-maydis Tehon & Daniels and *C. sorghi* var. *maydis* Ell. & Ev.

The conidia are straight to slightly curved, hyaline, slender, and multiseptate. The conidia of *C. zeae-maydis* are 40–165 × 4–9 μm with 6–12 septa (Fig. 26A), whereas the conidia of *C. sorghi* measure 30–300 × 2–5 μm, with 3–15 or more septa. A species of *Mycosphaerella* is believed to be the perfect stage of *C. zeae-maydis.*

Host Range

The causal fungi also attack barnyardgrass and Johnsongrass and other *Sorghum* spp.

Epidemiology

C. zeae-maydis is most prevalent within the USA in "no-till" or minimum tillage fields from Virginia to Missouri and surrounding states. It has also been reported from highlands of Mexico, Central America, Ecuador, and Peru.

Control

Hybrids differ greatly in their tolerance. Resistant hybrids are grown in the eastern USA.

Zonate Leaf Spot

This disease has been reported from the USA, Central and South America, and Africa. The lesions are reddish brown, water-soaked, and occur mainly on older leaves (Plate 22). They may enlarge up to 2.5–5 cm forming a zonate or targetlike pattern. Black sclerotia, 0.1–0.2 mm in diameter, may develop within necrotic tissue.

Causal Organism

Gloeocercospora sorghi D. Bain & Edg.

The conidia are borne in a pinkish to salmon-colored slimy matrix and are hyaline, needle-shaped, straight or slightly curved, and of variable length, 20–195 × 1.4–3.2 μm (av. 82.5 × 2.4 μm). The fruiting body is a sporodochium (Fig. 26B).

Host Range

The fungus also attacks sorghum, Sudangrass, Johnsongrass, sugarcane, and bentgrass.

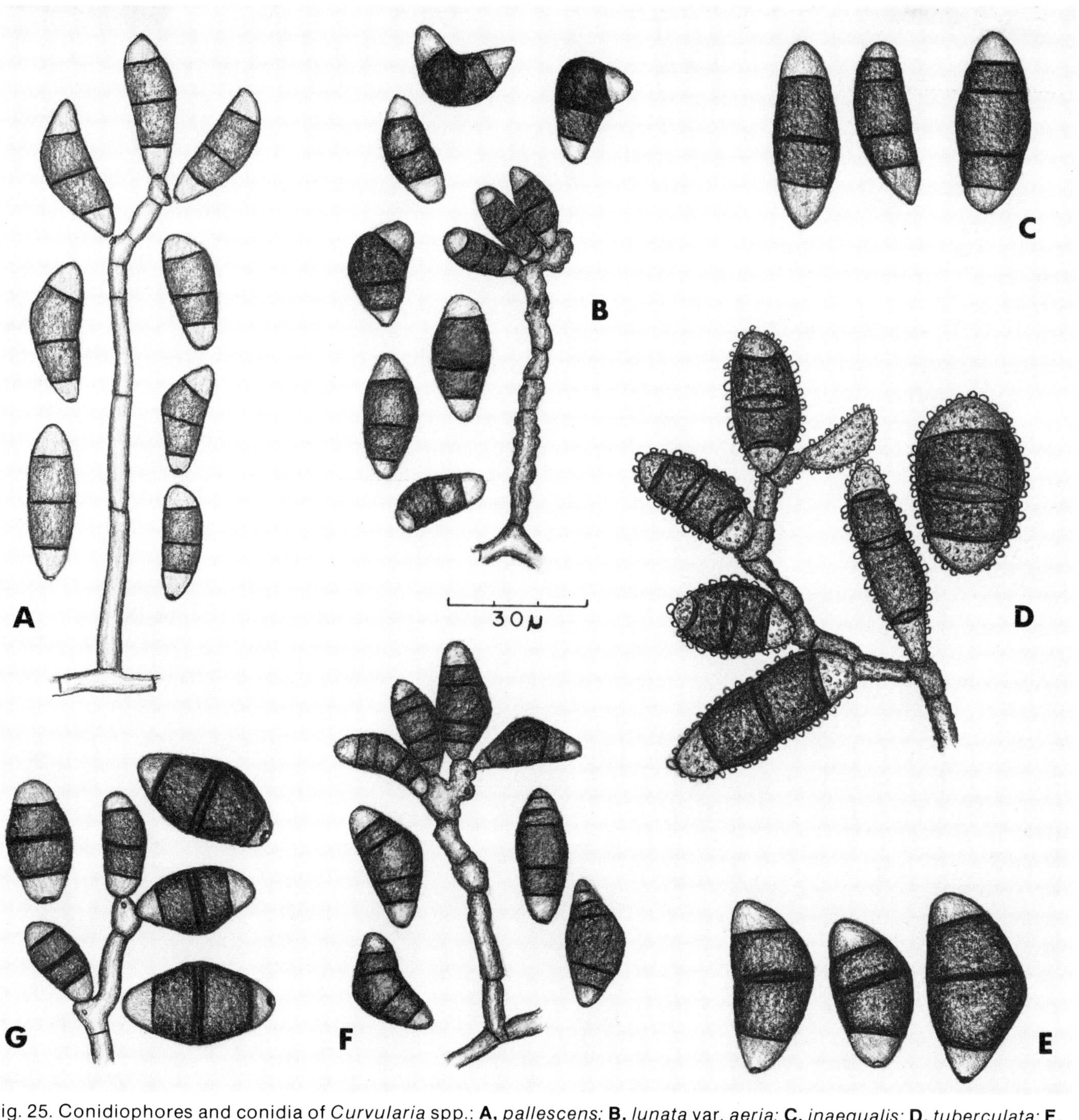

Fig. 25. Conidiophores and conidia of *Curvularia* spp.: **A**, *pallescens;* **B**, *lunata* var. *aeria;* **C**, *inaequalis;* **D**, *tuberculata;* **E**, *intermedia;* **F**, *senegalensis;* **G**, *eragrostidis* (after Ellis).

Ascochyta Leaf and Sheath Spots

Lesions are brown to reddish purple and ellipsoidal, becoming elongated and somewhat irregular with brown margins. Moderately abundant, subepidermal pycnidia (55–160 μm in diameter) sometimes develop in rows within the lesions. Spores of the causal fungi, *Ascochyta* spp., are one-septate, hyaline, ellipsoidal to irregular in shape, and rarely constricted at the septa (Fig. 26C and D).

A. zeae Stout—Conidia measure 8.5–13.5 × 3–4.5 μm; *A. maydis* Stout—Conidia measure 11–18 × 3–4.5 μm; *A. zeina* Sacc.—Conidia measure 18 × 7.5 μm.

Ascochyta leaf and sheath spots have been reported from the USA. Maize is apparently the only host.

Tar Spot

Black, sunken, glossy spots on the leaves are conspicuous when plants are heavily infected. Generally oval to circular lesions, 0.5–2.0 mm in diameter, may merge to form stripes up to 10 mm long (Plate 23). *Phyllachora maydis* Maubl., an obligate parasite, is the causal organism. The perithecia are nearly spherical, immersed in the mesophyll, and bear cylindrical asci, 80–100 × 8–10 μm in size, with short pedicels. Ascospores are broadly ellipsoidal, 9–12 × 5.5–7 μm in size, one-celled, fuliginous, and uniseriate in the asci (Fig. 27). The disease is most severe after pollination and may cause premature desiccation of the spike. Tar spot occurs in relatively cool, humid areas of the tropics where Northern leaf blight (page 16) is prevalent.

Didymella Leaf Spot

Didymella exitialis (Mor.) Müller causes small, elongate or elliptical leaf spots with brownish margins. The lesions enlarge and coalesce, forming white to cream-colored streaks or patches in which perithecia are sometimes formed.

Leptosphaeria Leaf Spot

Leptosphaeria sp. produces small tan-to-cream lesions on the leaves that become large and concentric or streaked, covering large areas of the leaves (Fig. 28). The disease is most conspicuous on the lower leaves at flowering time. The disease has been reported in the high humid areas of the Himalayas.

Septoria Leaf Blotch

In rather cool, humid environments *Septoria* sp. may produce small, light green to yellow-to-brown lesions on the leaves (Plate 24). The lesions may merge to produce severe blotching and necrosis. Large numbers of black pycnidia form in diseased tissue.

Banded Leaf and Sheath Spot

Rhizoctonia microsclerotia Matz, the imperfect state of *Corticium solani* (Prill & Delacr.) Bourd. & Galz. (Syn. *Thanatephorus cucumeris* (Frank) Donk and *Pellicularia*

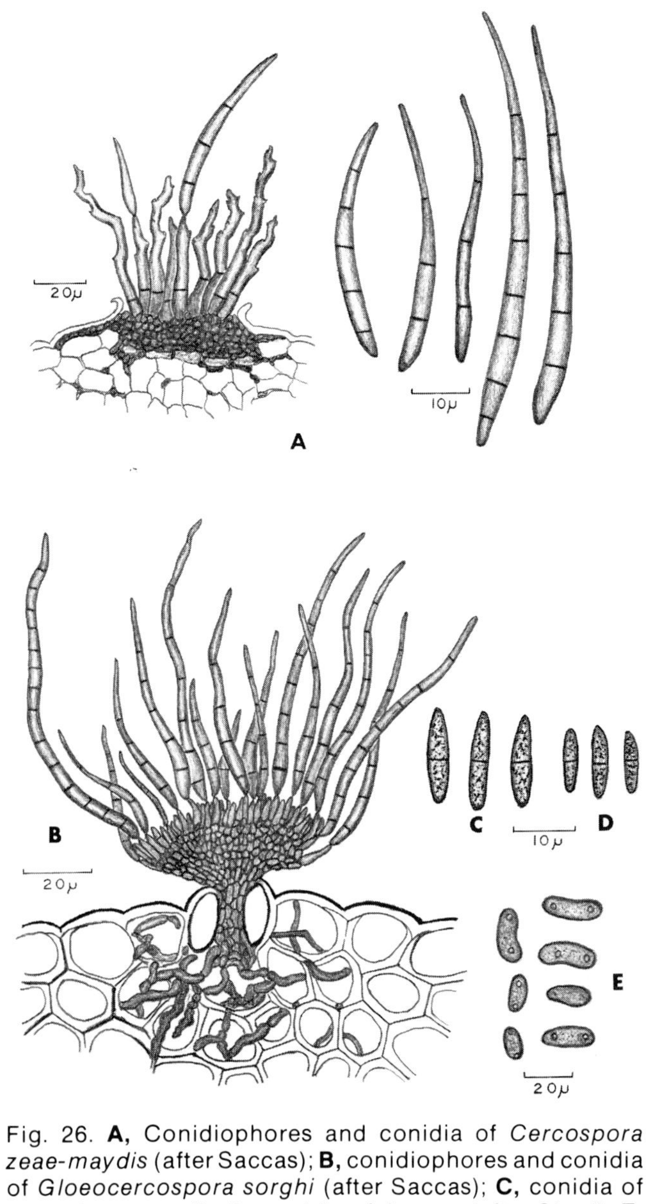

Fig. 26. **A,** Conidiophores and conidia of *Cercospora zeae-maydis* (after Saccas); **B,** conidiophores and conidia of *Gloeocercospora sorghi* (after Saccas); **C,** conidia of *Ascochyta maydis;* **D,** conidia of *A. zeae* (after Stout); **E,** conidia of *Phyllosticta maydis.*

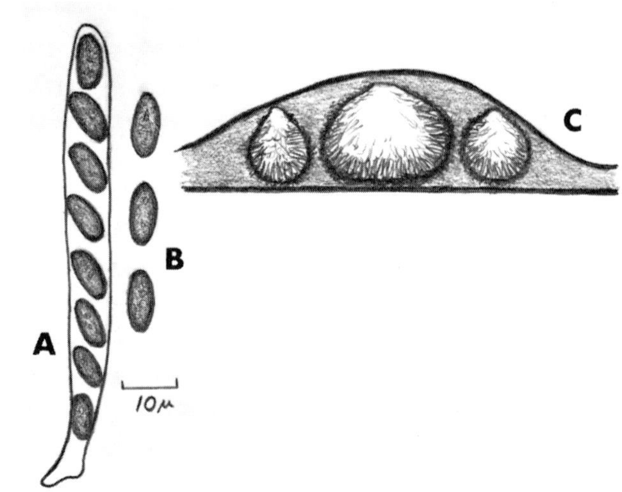

Fig. 27. *Phyllachora maydis.* **A,** An ascus containing eight ascospores; **B,** ascospores; **C,** cross-section of stroma showing three perithecia.

filamentosa (Pat.) Rogers), produces large, gray, tan, or brown discolored areas alternating with dark bands on infected leaves and leaf sheaths (Fig. 29). Brownish rotting of the ears may develop in the humid tropics. The mycelium is hyaline, septate, and branched at right angles. Sclerotia are spherical to oval in shape and brown to black. Mycelium on the ear husks appears pale brown and cottony (Plate 25). The disease is tropical or semitropical in distribution.

Diplodia Leaf Streak

In the USA, *Diplodia macrospora* Earle occurs mostly in the southern states as an ear-rotting (page 51) and stalk-rotting (page 44) pathogen, but in warm humid weather it can attack the leaves producing brown streaks, with characteristic yellow margins evident in the early stages. Leaf streaking is a common symptom in most Latin American countries, the streaks often extending almost the length of the leaf blade, emanating from the oval center of initial infection (Plate 26).

Hyalothyridium Leaf Spot

This disease is prevalent in several Latin American countries (e.g., Mexico, Costa Rica, Colombia), but has not been found in the USA. It occurs in humid areas at both low and high altitudes, and can be severe during the rainy season.

Symptoms

Leaf spots range from small, tan, elliptical spots with brown borders to large, almost circular, light brown blotches that enlarge up to 1.5 cm in diameter, sometimes so numerous that most of the leaf surface is covered (Plate 27). In some areas only the smaller lesions are found, and the cause of lesion enlargement and "rounding" in other areas is not known. In both types of lesions pycnidia bearing large hyaline muriform spores of a species of *Hyalothyridium* are found in close association with similar bodies bearing spermatial-like cells that may be part of the life cycle.

Causal Organism

Hyalothyridium sp.

Pycnidia are immersed, globose, thin-walled, and subhyaline to pale brown. The large circular ostioles open through stomates, becoming dark brown in old pycnidia. Conidia are hyaline, muriform, broadly oblong, and many celled with the cells often separating at maturity. Pycnidia measure 150–180 μm in diameter, with ostiole at maturity about 40 μm. Conidia are from 23 to 40 μm in length by 16–28 μm in diameter (mean 35 × 23 μm), each estimated to contain 10–15 cells.

Alternaria Leaf Blight

This disease has recently been reported from the central Corn Belt of the USA.

Symptoms

Chlorotic streaks, which later become necrotic, form on maize leaves of all ages. Leaves may be killed in seven days with 16–20 hr of leaf wetting each night.

Causal Organism

Alternaria alternata (Fr.) Keissler (Syn. *A. tenuis* C. G. Nees)

Colonies usually black or olivaceous black. The simple or branched conidiophores arise singly or in groups. They are straight or flexuous, sometimes geniculate, smooth, up to 50 μm long, and 3–6 μm thick with one or several conidial scars. Conidia form in long, often branched chains. Spores are obclavate, obpyriform, ovoid or ellipsoidal, often with a short beak up to one-third the length of the conidium. Conidia are pale to midgolden brown, smooth or verruculose, with up to eight transverse and commonly with several longitudinal or oblique septa. The spores are 20–63 μm long and 9–18 μm wide at the broadest part. The beak is pale and 2–5 μm thick. *A. alternata* is cosmopolitan, a very common saprophyte found on a wide range of plants and other substrata.

Fig. 28. Leptosphaeria leaf spot showing differences in lesion types. (Courtesy B. L. Renfro)

Fig. 29. Banded leaf and sheath spot. (Courtesy A. J. Ullstrup)

Epidemiology

Various forms of tissue damage, such as insect feeding scars and embedded sand grains, are apparently essential to infection that develops during periods of heavy dew.

Selected References

EDDINS, A. H. 1930. Dry rot of corn caused by *Diplodia macrospora* Earle. Phytopathology 20:439-448.

KAUR, R., M. M. PAYAK, and R. NATH. 1973. Curvularia leaf disease of maize. Kasetsart J. 7:76-80.

LATTERELL, F. M., and A. E. ROSSI. 1977. Further evidence for the genetic relationship between *Cercospora zeae-maydis* and a species of *Mycosphaerella*. Proc. Int. Mycol. Congr. II, Tampa, FL.

LATTERELL, F. M., and A. E. ROSSI. 1979. *Hyalothyridium* leaf spots of maize. (Abstr.). Phytopathology 69:536.

LATTERELL, F. M., A. E. ROSSI, and R. MORENO. 1976. A potentially serious pathogen of corn in the U.S.? Proc. Am. Phytopathol. Soc. 3, p. 228 (Abstr. 110).

MABADEJE, S. A. 1969. Curvularia leaf spot of maize. Br. Mycol. Soc. Trans. 52:267-271.

MANDOKHOT, A. M., and K. C. BASU CHAUDHARY. 1972. A new leaf spot of maize incited by *Curvularia clavata*. Neth. J. Plant Pathol. 78:65-68.

NELSON, R. R. 1956. A new disease of corn caused by *Curvularia maculans*. Plant Dis. Rep. 40:210-211.

RANE, M. S., M. M. PAYAK, and B. L. RENFRO. 1966. A Phaeosphaeria leaf spot of maize. Indian Phytopathol. Soc. Bull. 3:8-10.

SACCAS, A. M. 1952. Principaux champignons parasites du maïs (*Zea mays* L.) en Afrique Equatoriale Francaise. L'Agronomie Tropicale 7:5-42.

SCHIEBER, R., and A. S. MÜLLER. 1968. A leaf blight of corn (*Zea mays*) incited by *Fusarium moniliforme*. (Abstr.). Phytopathology 58:554.

STOUT, G. L. 1930. New fungi found on the Indian corn plant in Illinois. Mycologia 22:271-287.

TURNER, M. T., and KAREN BELL. 1978. *Diplodia macrospora* on leaves of corn from Tennessee. Plant Dis. Rep. 62:182.

ULLSTRUP, A. J. 1977. Diseases of corn. In: SPRAGUE, G. F. (ed.). Corn and Corn Improvement, Chap. 8. Am. Soc. Agron.: Madison, WI.

ULLSTRUP, A. J. 1978. Corn Diseases in the United States and Their Control. US Dep. Agric. Agric. Handb. 199. US Gov. Printing Office: Washington, DC.

Downy Mildews

Maize is attacked by at least nine species of three genera (*Peronosclerospora, Sclerophthora,* or *Sclerospora*). *Sclerophthora macrospora* and *Sclerospora graminicola* are found in most maize-growing areas of the world, but losses are seldom severe in the former and *S. graminicola* is rarely found on maize. On the other hand, *Sclerophthora rayssiae* var. *zeae, Peronosclerospora philippinensis, P. sacchari, P. maydis,* and *P. sorghi* cause yield losses of 50% or more in Southeast Asia, India, and Africa and deter increased maize production there. *P. sorghi* is established on maize, sorghum, and Johnsongrass in Central and South America and in certain southern and midwestern states in the USA. *P. spontanea* and *P. miscanthi* in the Far East are of little consequence on maize (Table I).

Symptoms caused by these fungi are somewhat similar and include chlorotic streaking, mottling, stunting, malformation of ears and tassels, and excessive tillering. Symptom expression is greatly affected by plant age, species of pathogen, and the environment.

Crazy Top

Crazy top occurs in most areas with temperate or warm-temperate climates. It appears rarely in tropical areas, but is widespread in the USA and has been reported in Canada, Mexico, eastern and southern Europe, Africa, and Asia. Crazy top causes substantial losses only in very localized areas.

Symptoms

Symptoms vary greatly with time of infection and degree of host colonization by the fungus. Generally, excessive tillering (6 to 10 tillers per plant), rolling, and twisting of the upper leaves appear first (Fig. 30A). The most characteristic symptom is the partial or complete proliferation of the tassel, which continues until the tassel resembles a mass of leafy structures (Fig. 30B). These modified leaflike influorescences are described as "crazy top." Phyllody may also occur in the ears. Leaves of severely infected plants are often narrow, straplike, and leathery (Fig. 31). Stunting and chlorotic striping of leaves are other common symptoms.

Causal Organism

Sclerophthora macrospora (Sacc.) Thirum., Shaw & Naras. (Syn. *Sclerospora macrospora* Sacc.; *Phytophthora macrospora* (Sacc.) Ito & Tanaka)

Sporangia are hyaline, lemon-shaped, operculate, 60–100 × 30–65 μm, and attached to short, simple, hyphoid sporangiophores emerging from stomata. The sporangia germinate by the release of numerous biciliate, subspherical to reniform, hyaline zoospores or occasionally by germ tubes (Fig. 32A). After a period of motility, the zoospores encyst and produce germ tubes. The oospores are hyaline to yellowish, multinucleate, somewhat globose, and 45–75 μm in diameter, with granular contents. They germinate by a thin-walled tube bearing a sporangium, which in turn produces zoospores. The mycelium is coenocytic, thin-walled, intercellular, and abundant in meristematic tissues.

Histopathology

The mycelium of *S. macrospora* can be detected in brace roots, stalk tissue, leaves, and floral structures of affected plants by staining with zinc chloriodide. The mycelium stains a deep purplish-blue while the host tissue remains nearly colorless. The oogonia and oospores are situated in "nests" of several spores in vascular bundles or in their parenchymatous sheath cells within straplike leaves and leaf sheaths of stunted plants.

Disease Cycle

The oospores germinate in saturated soil to produce sporangia from which zoospores escape that penetrate host tissues. After a short period of motility, they encyst and each produces a germ tube. The mycelium develops systemically and is most abundant in meristematic tissues. The sporangia are borne on sporangiophores projecting from stomata in the leaves, but these are produced sparingly on maize. Oogonia are produced abundantly in infected tissues.

Epidemiology

Unlike other sclerophthoras, the crazy top fungus develops systemically where soils have been flooded shortly after planting or before plants are in the four-leaf to five-leaf stage. Saturation for 24–48 hr is sufficient for infection to occur, probably by initiating germination of soilborne oospores and providing liquid for movement of zoospores to infection sites on maize. Stricken nearby collateral hosts (particularly wild grasses) sometimes provide sporangial inoculum. Infection happens in a wide range of soil temperatures. The optimum temperature for sporulation on excised leaves is 24–28°C; for sporangial germination it is 12–16°C (22–25°C for finger millet in India). Little is known about physiologic specialization of parasitism in *S. macrospora*. Seed transmission of the pathogen has been demonstrated but is generally considered unimportant in dissemination of the fungus.

S. macrospora has been reported on more than 140 species of the Gramineae. Hosts include oats, wheats, sorghums, rice, crabgrass, witchgrass, green foxtail, St. Augustine grass, and barnyardgrass. On finger millet in India, it causes considerable losses.

Control

Provide adequate soil drainage or avoid planting in low, wet spots.

Sorghum Downy Mildew

This disease is present in the USA, many Latin American countries, Southeast Asia, India, Israel, Italy, and Africa.

Symptoms

Systemically infected plants are chlorotic, sometimes stunted, and occasionally have white-striped leaves and abnormal seed set (Fig. 33A). The chlorotic area of a leaf always includes the base of the blade, with the transverse margin usually sharply defined between the

Table I. Downy Mildews of Maize

Scientific Name (Common Name)	Sporangiophores or Conidiophores[a]	Sporangia or Conidia	Oospores	Geographical Distribution	Hosts other than Maize
Sclerophthora macrospora (crazy top)	Hyaline, short (av. 13.8 μm), simple, hyphoid, determinate	Sporangia hyaline, lemon-shaped, operculate, 60–100 × 30–65 μm; germinate by numerous zoospores	Hyaline—pale yellow, develop mainly in vascular bundles, 45–75 μm in diameter; germinate by sporangia	North America, Mexico, South America, Europe, Africa, Asia	*Avena sativa*, *Digitaria* spp., *Panicum* spp., *Setaria* spp., *Sorghum* spp., *Triticum* spp.
S. rayssiae var. *zeae* (brown stripe)	Hyaline, short, hyphoid, determinate	Sporangia hyaline, ovate to almost cylindrical, operculate, 29–66.5 × 18.5–26 μm; germinate by zoospores	Brown, spherical, 29.5–37 μm in diameter; germinate by sporangia	India, Nepal, Thailand	*Digitaria* spp.
Sclerospora graminicola (Graminicola downy mildew)	Hyaline, long (av. 268 μm), bloated, nonseptate, irregularly dichotomously branched; ephemeral	Sporangia hyaline, broadly elliptical, operculate, and with papillate, 14–23 × 11–17 μm; germinate by zoospores	Pale brown, spherical, usually smooth-walled, 35 μm in diameter; said to germinate by germ tube or by sporangia	Worldwide, but rare on maize and only in Israel and USA	*Euchlaena* spp., *Panicum* spp., *Pennisetum* spp., *Saccharum* spp., *Setaria* spp., *Sorghum* spp.
Peronosclerospora sorghi (sorghum downy mildew)	Hyaline, 180–300 μm long, bloated, usually dichotomously branched 2–3 times and septate near base; ephemeral	Conidia hyaline, oval to almost spherical, 15–26.9 × 15–28.9 μm; germinate by germ tube	Usually brown to subhyaline, spherical, 25–42.9 μm in diameter; germinate by wide germ tube	North America, Latin America, Southeast Asia, India, Africa, Italy(?)	*Euchlaena* spp., *Heteropogon contortus*, *Panicum* spp., *Pennisetum* spp., *Sorghum* spp.
P. philippinensis (Philippine downy mildew)	Hyaline, long (150–400 μm), bloated, widening abruptly; dichotomously branched 2–4 times; ephemeral	Conidia hyaline, elongate-ovoid to round-cylindrical, 27–39 × 17–21 μm and slightly rounded at apex; germinate by germ tube	Rare; spherical, smooth-walled, 22.6 μm in diameter; germinate by side germ tube	Philippines, Indonesia, India, Nepal, Africa(?)	*Avena sativa*, *Euchlaena* spp., *Zea mays mexicana* × *Z. mays* hybrids, *Saccharum* spp., *Sorghum* spp.
P. sacchari (sugarcane downy mildew)	Hyaline, 160–170 μm, bloated, widening gradually, dichotomously branched 2–3 times; ephemeral	Conidia hyaline, elliptical, oblong to conical, 25–41 × 15–23 μm with round apex; germinate by germ tube	Yellow to yellow-brown, globular to slightly angular; 40–50 μm in diameter; germinate by germ tube	Australia, Fiji Islands, New Guinea, India, Philippines, Taiwan, Thailand	*Euchlaena* spp., *Saccharum* spp., *Sorghum* spp.
P. maydis (Java downy mildew)	Hyaline, long (150–550 μm), bloated; dichotomously branched 2–4 times; ephemeral	Conidia hyaline, spherical to subspherical, 27–39 × 17–23 μm; germinate by germ tube	Unknown	Indonesia	*Zea mays* mexicana × *Z. mays* hybrids

[a] Conidiophore and sporangiophore length and branching in species of *Sclerospora* and *Peronosclerospora* are variable, depending somewhat on ambient air conditions during their formation.

diseased and healthy tissue appearing further up the blade in successively formed leaves ("half-leaf symptom"), until the whole blade is chlorotic. Leaves of infected plants tend to be narrower and more erect than those of healthy plants. A white downy growth may appear on both surfaces of infected leaves; in some regions diseased plants may have phyllodied tassels (Fig. 34). Sorghum leaves may shred due to oospore formation (Fig. 33B). Plants of tolerant varieties may show symptoms of systemic infection but have normal seed production. Other plants may be barren. Long, narrow, chlorotic, local lesions develop on some maize lines. Ear shanks may be elongated. Oospores, if present, tend to appear more in the ear bracts and leaves within the striped areas.

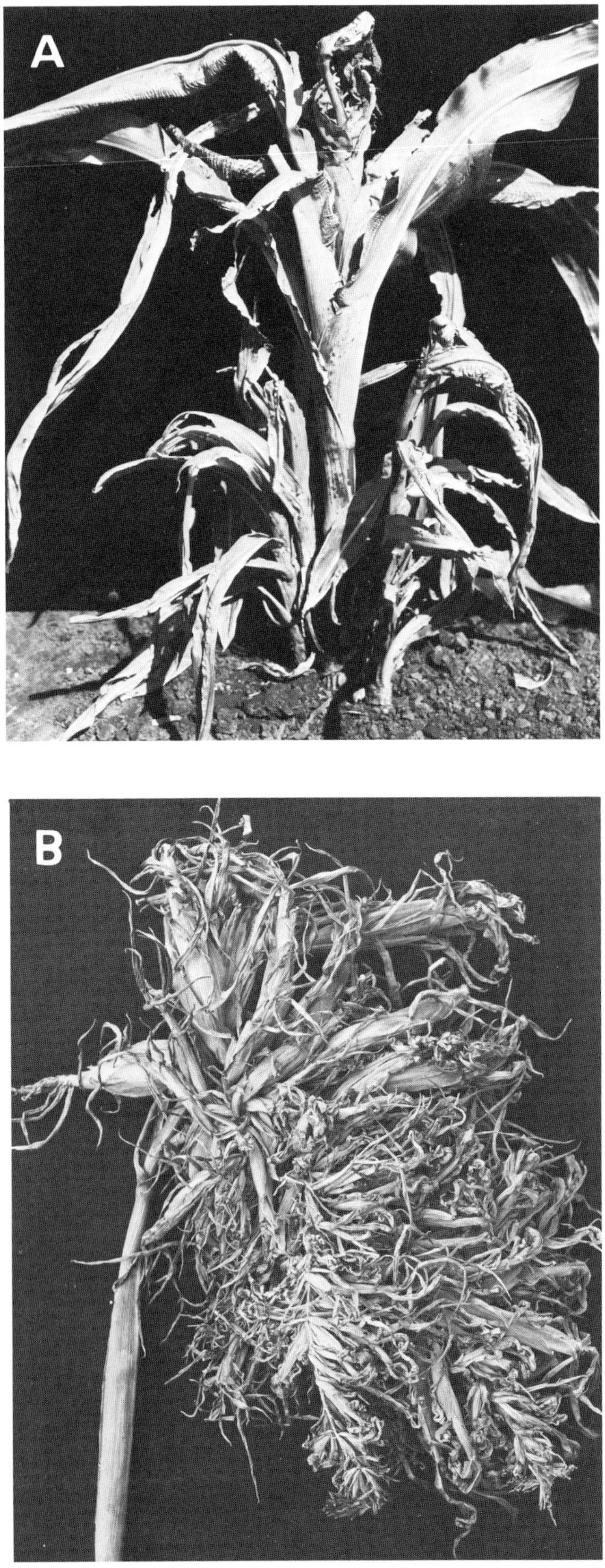

Fig. 30. Crazy top. **A,** Excessive tillering, rolling, and twisting of leaves; **B,** tassel has been replaced by a mass of leafy structures.

Causal Organism

Peronosclerospora sorghi (Weston & Uppal) C. G. Shaw (Syn. *Sclerospora sorghi* Weston & Uppal, *S. graminicola* var. *andropogonis-sorghi* Kulk.)

The conidiophores are erect, spreading, fragile, hyaline, 180–300 μm long, usually dichotomously branched, and emerge singly or in groups from stomata on the lower and sometimes the upper sides of leaves. The conidia are hyaline, oval to subspherical, 14.4–27.3 × 15–28.9 μm, and are borne on elongated, tapered sterigmata about 13 μm long (Fig. 32B). They germinate by germ tube in free water and penetrate corn leaves

Fig. 31. Crazy top. Severe infection showing complete suppression of tassels and ears and narrow, straplike leaves. (USDA photo)

directly through the stomata. The oogonia are spherical, 40–55 μm in diameter, and embedded among mesophyll cells between fibrovascular bundles. The oospores measure 25–42.9 μm (av. 36 μm), are hyaline and spherical, with light yellow outer walls.

Disease Cycle

The oospores of *P. sorghi,* which survive several seasons in soil, and conidia probably constitute the primary inoculum. The oospores germinate by germ tubes and attack underground parts of susceptible maize seedlings, which become systemically infected. No infection by oospores occurs if seedlings emerge in cool soil. Conidia on the leaves are windborne and provide secondary inoculum which can induce systemic infection in plants up to four weeks old. Oospores of *P. sorghi* are produced less frequently and abundantly on maize than on sorghums, and appear in both hosts only in systemically infected plants. Shredding of leaves occurs only in sorghums (Fig. 33B). Local lesions induced by conidia appear to a much greater degree in maize than in sorghums, where they do damage.

Epidemiology

Conidia are produced between 17 and 29°C (optimum often 24–26°C depending on the geographic isolate). Conidial germination requires a saturated atmosphere, and moderate temperatures of 21–25°C are optimal for some isolates; 15°C for others. The conidia are delicate and lose their viability after 3–4 hr. High levels of systemic infection can occur between 11 and 32°C with a wet period of 4 hr or longer. *P. sorghi* can also infect sorghum, Johnsongrass, and teosinte. In India, there is evidence of physiologic specialization in *P. sorghi* with maize and sorghum pathotypes described; only the former can infect tanglehead and oospores form only in that grass.

Control

1. Resistant hybrids and varieties offer the best means of control.
2. Destroy infected crop debris and *Sorghum* weeds.
3. Rogue and destroy infected plants as they appear in the field.
4. Protective and systemic fungicide seed treatment.
5. Plant at the onset of the rainy season (India, Nepal) in cool soil.
6. Avoid maize-sorghum rotations in fields where the disease has occurred, and do not sow maize adjacent to a field of maize or sorghum that can provide conidia for infection.

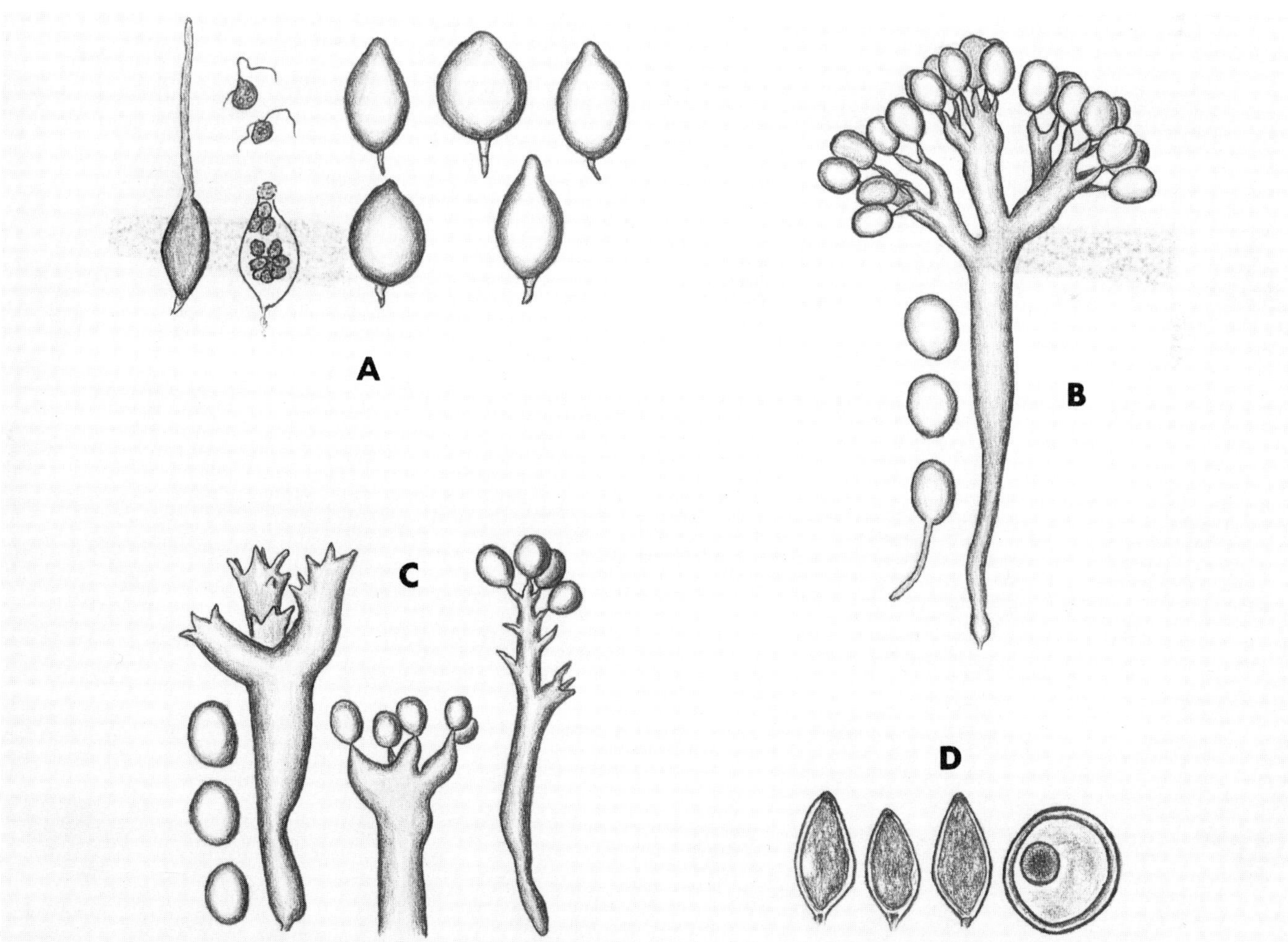

Fig. 32. Four species of downy mildew fungi: **A,** Typical sporangia of *Sclerophthora macrospora* showing germination by germ tube and by release of zoospores; **B,** sporangiophore and sporangia of *Peronosclerospora sorghi* (one sporangium is putting forth a germ tube); **C,** sporangiophores and sporangia of *Sclerospora graminicola;* **D,** three sporangia and an oospore of *Scleropthora rayssiae* var. *zeae.* Note well-developed peduncles on sporangia and prominent oil globule in oospore.

Graminicola Downy Mildew

This disease, also called "green ear," occurs on various grasses throughout the world. It has been found in maize only in Israel and the USA and is rare in both countries. In the USA it passes from *Setaria*; in Israel from pearl millet.

Symptoms

Grayish blotching and mottling of leaves with a white, arachnoid to downy growth developing on discolored areas. Chlorotic streaking or striping, as well as stunting, are common symptoms. Leaves may be thick, corrugated, and fragile. Symptoms usually appear 10 days after the plumule emerges. Affected plants will occasionally outgrow the disease.

Causal Organism

Sclerospora graminicola (Sacc.) Schroet.

Sporangiophores are long (av. 268 μm), nonseptate, and stalklike with a slightly bulbous foot. Sporangia are borne on short sterigmata, two to six on a branch, 14–23 × 11–17 μm, with a prominent papillate operculum at the apex. They are broadly elliptical (Fig. 32C). Each sporangium germinates by releasing three or more irregularly kidney-shaped, biciliate zoospores 9–12 μm in diameter. Sporangia are reported to germinate occasionally by germ tubes, but these reports probably result from germ tube formation from an encysted zoospore trapped within the sporangium. The oospores are pale brown, 22.5–35 μm in diameter, and germinate by germ tubes. The oogonia are 30–60 μm in diameter, with reddish-brown or amber-brown, heavy, uneven outer walls, 4–12 μm thick.

Disease Cycle

Primary infection occurs from oospores that overwinter in the soil. In India, they may survive in soil for at least three years; some remain viable after passage through the alimentary canal in cattle. The oospores may also be seedborne. The mode of germination is still uncertain; they systemically infect maize seedlings. Disease development and spread are

Fig. 33. Sorghum downy mildew. **A,** Systemically infected plant; **B,** shredded leaves due to oospore formation. (Courtesy R. A. Fredriksen)

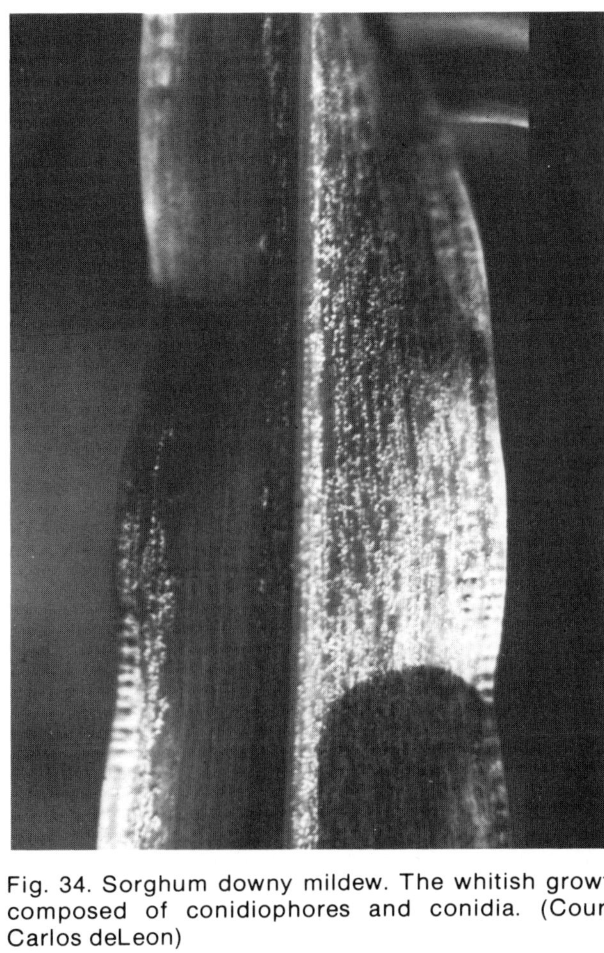

Fig. 34. Sorghum downy mildew. The whitish growth is composed of conidiophores and conidia. (Courtesy Carlos deLeon)

favored by light rains or heavy dews and moderate night temperatures, optimal for production of sporangia. Secondary spread to other hosts is by windborne sporangia produced superficially, but only where free moisture on plants is sustained.

Epidemiology

Temperatures of 24–32° C for two days after sowing are conducive to heavy infection. The optimal temperature for sporangial formation and germination is 17°C with the maximum at 34°C. Species of *Echinochloa, Euchlaena, Holcus, Panicum, Pennisetum, Saccharum, Setaria,* and *Sorghum* are susceptible to natural or artificial inoculation. Heavy losses are encountered in pearl and foxtail millets. Strong physiologic specialization exists with most isolates from pearl millet not attacking *Setaria* spp. and vice versa.

Control

None needed for maize.

Brown Stripe Downy Mildew

This local lesion disease can be very destructive in India and has been reported from most maize-growing areas of that country, as well as from Nepal and Thailand.

Symptoms

Initially, lesions develop on the leaves as narrow, chlorotic or yellowish stripes, 3–7 mm wide, with well-defined margins and delimited by the veins (Plate 28). The stripes later become reddish to purple. Lateral development of lesions causes severe striping and blotching. Seed development may be suppressed and the plant may die prematurely if blotching occurs prior to flowering. Sporangia on the leaves appear as downy or woolly growth on both surfaces of the lesions. Floral or vegetative parts are not malformed, and the leaves do not shred.

Causal Organism

Sclerophthora rayssiae Kenneth, Kaltin, and Wahl *var. zeae* Payak & Renfro

The oogonia are subglobose, thin-walled, hyaline to light straw-colored, and 33–44.5 μm in diameter. The oospores are spherical to subspherical, 29.5–37 μm in diameter, with smooth, glistening walls 4 μm thick, and confluent with the oogonial wall. The oospores are centrally located within the oogonia and have hyaline contents including a prominent oil globule. Oospores and oogonia are numerous and scattered in the leaf mesophyll or under stomata. Sporangiophores are short, determinate, and arise from hyphae in the substomatal spaces. The sporangia are produced sympodially in groups of two to six, are hyaline, ovate to cylindrical with a prominent peduncle, smooth-walled, 29.0–66.5 × 18.5–26.0 μm and individually produce four to eight zoospores on germination (Fig. 32D). The encysted zoospores are hyaline, spherical, and 7.5–11 μm in diameter.

Disease Cycle

S. rayssiae overseasons as oospores in infected debris in the soil. On germination, the oospore produces a sporangiophore bearing a sporangium that liberates the zoospores that constitute the primary inoculum. In the presence of enough moisture or high temperatures, the sporangium may produce a germ tube that can also infect maize leaves. The pathogen may also overseason in grass hosts as mycelium from which sporangia are produced. This is another source of inoculum. Secondary spread is by sporangia produced after primary infection is established.

Epidemiology

The severity of brown stripe downy mildew is influenced by temperature and moisture. Sporangia are produced at low temperatures (20–22° C), while oospores are formed at higher temperatures. Soil temperatures of 28–32° C favor disease development. Sporangial production, germination, and infection require a film of moisture for 12–96 hr. Zoospore germination occurs over a range of temperatures (15–30° C), with an optimum at 22–25° C. Sporangia and spores of the pathogen are disseminated by wind, rain, and animals. The pathogen infects crabgrass. Zinc deficiency predisposes plants to infection.

Control

1. Cultural practices; plant before the rainy season begins.
2. Resistant hybrids and varieties.
3. Fungicides are promising as foliar sprays after symptoms appear or applied as a soil drench.

Sugarcane Downy Mildew

Severe outbreaks of this disease on corn have limited maize production in Taiwan. Yield losses of 30 to 60% have been reported. The disease appears to be confined to tropical and subtropical regions and has been reported from Australia, Fiji Islands, New Guinea, India, The Philippines, Taiwan, and possibly Thailand. It is a potential threat to maize production, particularly in Asia.

Symptoms

Sugarcane downy mildew is characterized by local lesions leading to systemic infection. Initial lesions are small, round, chlorotic spots on the leaves appearing two to four days after infection. Systemic symptoms appear as pale yellow to white stripes or streaks at the base of the third to sixth oldest leaves. Several streaks may form in each leaf and may extend the length of the leaf (Fig. 35). On the leaves of some varieties, or on older leaves, the streaks may be narrow and discontinuous. On late-infected or mildly infected plants, streaks may disappear as plants approach maturity. White, downy, or powdery masses of conidia and conidiophores appear on both leaf surfaces as well as the leaf sheaths and husks (Plate 29). This downy growth is usually produced at night under moderate temperatures (25° C), especially when dew is present. Plants may be distorted with small, numerous, poorly filled ears and improperly formed tassels. Ear shanks may be elongated.

Causal Organism

Peronosclerospora sacchari (T. Miyake) C. G. Shaw. (Syn. *Sclerospora sacchari* T. Miyake)

Conidia and conidiophores are superficial. The conidiophores are erect, hyaline, 160–170 μm long, arising singly or in pairs from stomata. The conidia are hyaline, elliptical, oblong or conical, 25–41 × 15–23 μm, with a round apex and thin, smooth walls (Fig. 36A). The oospores are globular or slightly angular, yellow, 40–50 μm in diameter, with walls 3.8–5 μm thick. The conidia and oospores germinate by hyaline, slender, nonseptate germ tubes, 3.8 μm wide.

Disease Cycle

The fungus overseasons as mycelium in sugarcane. Under favorable conditions, germ tubes from conidia penetrate maize leaves through stomata forming tiny, round lesions covered with ectophytic mycelia at infection sites. The mycelium spreads intercellularly by inserting knoblike haustoria into host cells. After the endophytic mycelium invades tissues of the growing point, the first streaks appear on young leaves. Once established in maize seedlings, the fungus produces large numbers of sporangia that serve as secondary inocula for maize or sugarcane seedlings. The role of oospores in infection under natural conditions has not been established.

Epidemiology

Moisture, temperature, and age of plants are the important factors in disease development. The optimum temperature for conidial formation, germination, and infection is 20–25°C in the presence of free water. Sporulation is nocturnal and decreases with increasing light intensity. Plants less than one month old are very susceptible to infection, becoming resistant with age. Conidia are windborne but are viable for only a few hours even under optimum conditions. Other hosts of *P. sacchari* include sugarcane, teosinte, gamagrass, broomcorn, and grain sorghum.

Control

1. Resistant hybrids and varieties. Due primarily to the development of resistant maize and sugarcane, this disease is no longer of consequence in Australia and Thailand.

2. Cultural practices: Prevention—eliminate the fungus in sugarcane by using healthy sugarcane sets; isolation—grow maize in disease-free areas where sugarcane is not grown extensively, and do *not* interplant with sugarcane; eradication—rogue and destroy diseased plants; fungicides—systemic seed treatment and sprays are a general practice in some areas.

Philippine Downy Mildew

This is the most serious downy mildew disease in The Philippines where yield losses range from 15 to 40% or more. It also occurs in India, Indonesia, Nepal, and possibly in Africa.

Symptoms

Systemic symptoms may appear in the first true leaf as a complete chlorosis or chlorotic stripes nine days after planting while local symptoms may appear when plants are in the two-leaf to three-leaf stage until tassels and silks are formed. In general, infected leaves show long, chlorotic streaks with a downy growth of conidia and conidiophores. Tassels may be malformed producing less pollen while ears may be aborted, resulting in partial or complete sterility. If infection occurs early, plants are stunted and may die (Plate 30).

Causal Organism

Peronosclerospora philippinensis (Weston) C. G.

Fig. 35. Sugarcane downy mildew. Note the long, pale yellow to white stripes on the leaves caused by *Peronosclerospora sacchari*. (Courtesy B. L. Renfro)

Shaw (Syn. *Sclerospora philippinensis* Weston)

The mycelium is branched, slender (8 μm in diameter), irregularly constricted, and inflated. The haustoria are simple, vesiculiform to subdigitate, and small (8×2 μm). Erect conidiophores grow out of stomata, are 150–400 $\times$ 15–26 μm, and dichotomously branched two to four times. Branches are robust, sterigmata are ovoid to subulate, slightly curved, and 10 μm long. The conidia are elongate-ovoid to round-cylindrical, hyaline, 27–39 $\times$ 17–21 μm, and slightly rounded at the apex (Fig. 36B).

Disease Cycle

P. philippinensis penetrates maize leaves through stomata by germ tubes from germinating conidia. Penetration is followed by intercellular invasion of the mesophyll cells. The fungus then grows through the leaf sheath into the stem becoming established in the shoot apex where it persists. Invasion of the shoot apex produces chlorotic streaks on the lower leaves and, later on, the younger leaves. The pathogen does not invade roots. The fungus is seedborne and may result in local lesions leading to systemic infection. Conidia and conidiophores are produced on chlorotic areas on the leaf blades and sheaths. Wind and splashing water are the main agents of conidial dispersal. Four-week-old plants became highly resistant.

Epidemiology

Night temperatures of 21–26° C and free moisture are essential for conidial production, germination, and infection. The optimum temperature for spore germination is 19–20° C. Physiologic races of the fungus exist. The host range of *P. philippinensis* includes oats (by inoculation), teosinte, sugarcane, sorghum, Johnsongrass, and *Sorghum propinquum*. Some of these hosts may be reservoirs of inoculum.

Control

1. Resistant hybrids and varieties.
2. Sanitation; rogue and destroy infected plants.
3. Systemic fungicide seed treatment and fungicide-oil sprays are applied in some areas.

Java Downy Mildew

This systemic downy mildew is of great importance in Indonesia where reductions in yield of 40% may occur. Late-planted, early rainy-season maize is most severely damaged, especially if overfertilized with N or planted after maize or sugarcane.

Symptoms

White-to-yellow streaking (Fig. 37), followed by necrosis and browning of the streaks, are characteristic leaf symptoms. The fungus may develop systemically causing severe chlorosis in the upper leaves. Infected plants may be stunted, sterile, and often lodge. Downy growth on the chlorotic streaks is common. Plants older than four weeks are highly resistant to infection.

Causal Organism

Peronosclerospora maydis (Racib.) C. G. Shaw (Syn. *Sclerospora maydis* (Racib.) Butl.)

The pathogen produces two kinds of hyphae—one is straight and sparsely branched; the other lobed, irregularly branched, and clustered. Conidiophores arise from stomata, are dichotomously branched two to four times above, and the branches are robust and 150–550 μm long with basal cells 60–180 μm long. The conidia are hyaline, spherical to subspherical, and 27–39 $\times$ 17–23 μm (Fig. 36C). Oospores are unknown.

Disease Cycle

Infected maize plants grown during the dry season are the primary source of inoculum in Indonesia. The fungus may survive as mycelium in kernels, but this is a minor source of inoculum. Infection by conidia occurs through stomata of young plants and local lesions elongate toward the meristem, inducing systemic infection. Conidia and conidiophores are produced superficially on wet leaves in the dark. If infection arises from seed, the cotyledonary leaf is always infected, but if inoculum is from oospores, the cotyledonary leaf is never infected.

Epidemiology

Conidial formation in *P. maydis* requires free moisture, darkness, and temperatures below 24°C. Conidia germinate under these conditions to produce germ tubes. Teosinte and teosinte $\times$ maize hybrids are hosts of this pathogen but are unimportant in its survival. No collateral wild grass hosts are known.

Control

1. Resistant hybrids and varieties.
2. Systemic fungicide seed treatments.

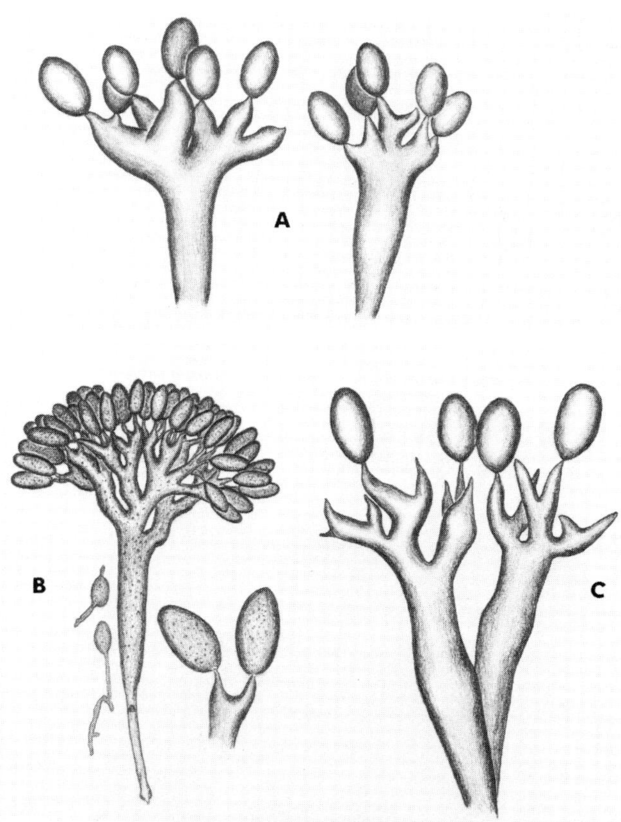

Fig. 36. Three species of downy mildew fungi: **A,** Sporangiophores and sporangia of *Peronosclerospora sacchari;* **B,** sporangiophores and sporangia of *P. philippinensis* including close-ups of two sporangia and their sterigmata and two sporangia germinating by germ tube; **C,** sporangiophores and sporangia of *P. maydis.*

Minor Downy Mildews

Spontaneum Downy Mildew

This disease was reported to be serious on corn about 1920 in the Viscayan Islands of The Philippines. In recent years, it has been rare on maize in this country and has not been reported elsewhere on corn except in Thailand. Other hosts include wild and cultivated sugarcane, teosinte, and a wild grass (*Micanthus japonicus*).

Causal Organism

Peronosclerospora spontanea (Weston) C. G. Shaw (Syn. *Sclerospora spontanea* Weston)

The conidiophores are erect, hyaline, 315.6–441 μm long, emerging singly or in groups from stomata on the lower leaf surface. Each is comprised of a basal cell, a main axis, and a more-or-less complex dichotomously branched system. The main axis is usually expanded above the septum to 22.7–28.6 μm and constricted below the branches to about 21 μm. The conidia are hyaline, elongate, ellipsoidal or cylindrical with rounded apices, lack a papilla, and are borne on slender straight sterigmata about 13 μm long. Most conidia are 15.5–18.1 × 38.8–49.2 μm as opposed to those of *P. sorghi* (14.4–27.3 × 15–28.9 μm). The oogonia are amber brown, usually irregularly developed with several folds in the wall, ranging from 55.1 to 82.1 μm; the wall thickness is variable. The spherical oospores are mostly 39.2–53.3 μm in diameter (mean 45.4–48.8 μm). The spore wall is hyaline to pale amber and 1.8–6.1 μm thick. Oospores were not seen in corn tissue.

Peronosclerospora miscanthi

Peronosclerospora miscanthi (T. Miyake) C.G. Shaw (Syn. *Sclerospora miscanthi* T. Miyake) has been used to inoculate maize but has not been found occurring naturally on this host. It attacks *Miscanthis japonica* in Taiwan.

Selected References

CONFERENCE ON THE DOWNY MILDEW DISEASES OF MAIZE. 1976. Kasetsart J. 10:79-207.

FREDERIKSEN, R. A. et al. 1973. Sorghum downy mildew: A disease of maize and sorghum. Tex. Agric. Exp. Stn. Res. Monograph 2.

FREDERIKSEN, R. A., and B. L. RENFRO. 1977. Global status of maize downy mildew. Annu. Rev. Phytopathol. 15:249-275.

KENNETH, R. G. 1975. *Sclerospora graminicola, Sclerospora sorghi.* Set 46, No. 15, 451 and 452. Commonw. Mycol. Inst. Descr. of Pathogenic Fungi and Bacteria.

PAYAK, M. M., and B. L. RENFRO. 1967. A new downy mildew disease of maize. Phytopathology 57:394-397.

PAYAK, M. M., B. L. RENFRO, and S. LAL. 1970. Downy mildew diseases incited by *Sclerophthora*. Indian Phytopathol. 23:183-193.

PUPIPAT, U. 1974. Host range, geographic distribution and physiologic races of the maize downy mildews. Trop. Agric. Res. (Jpn.) Ser. 8:63-80.

SHARMA, R. C., M. M. PAYAK, B. K. MUKHERJEE, and J. LILARAMI. 1976. Multiple disease resistance in maize. Kasetsart J. 10:125-136.

SHAW, C. G. 1974. The taxonomy of graminicolous downy mildews, with emphasis on those attacking maize. Trop. Agric. Res. (Jpn.) Ser. 8:47-55.

SHAW, C. G. 1978. *Peronospora* spp. and other downy mildews of the Gramineae. Mycologia 70:594-604.

SCHMITT, C. G., and R. E. FREYTAG. 1977. Response of selected resistant maize genotypes to three species of *Sclerospora*. Plant Dis. Rep. 61:478-481.

SYMPOSIUM ON DOWNY MILDEW OF MAIZE. 1975. Trop.

Fig. 37. Java downy mildew. Long, white to yellow streaks on the leaves, caused by *Peronosclerospora maydis,* later turn brown. (Courtesy B. L. Renfro)

Agric. Res. (Jpn.) Ser. 8.
TITATARN, S., and R. SYAMANANDA. 1978. The occurrence of *Sclerospora spontanea* on *Saccharum spontaneum* in Thailand. Plant Dis. Rep. 62:29-31.
ULLSTRUP, A. J. 1970. Crazy top of maize. Indian Phytopathol. 23:250-261.
WATERHOUSE, GRACE M. 1964. The genus *Sclerospora*. Diagnosis (or descriptions) from the original papers and a key. Commonw. Mycol. Inst. Misc. Publ. 17.
WESTON, W. H. JR. 1920. Philippine downy mildew of maize. J. Agric. Res. 19:97-122.
WORKSHOP ON THE DOWNY MILDEW DISEASES OF MAIZE AND SORGHUM. 1970. Indian Phytopathol. 23:173-435.

Ergot

This disease, also known as horse's tooth (diente de caballo), occurs in certain humid valleys of central Mexico.

Symptoms

Sclerotia that replace the kernels are first white to cream-colored, 5–8 × 2–5 cm in size, soft, sticky, and hollow. Each ear may contain one to several sclerotia. Honeydew of the sclerotia is filled with small, hyaline, ovoid spores. Mature sclerotia are comma-shaped (resembling a horse's tooth) and white to grayish brown. Those beneath the husks are lighter in color (Fig. 38).

Causal Organism

Claviceps gigantea Fuentes, de la Isla, Ullstrup & Rodriguez (imperfect state *Sphacelia* sp.)

The sclerotia are white-to-cream becoming grayish brown with age and have pink to lavender centers. The perithecia are flask-shaped and arranged peripherally in stromatic heads. The asci in the perithecia are cylindrical and hyaline, and the ascospores are filiform, hyaline, nonseptate, and 176–186 × 1.5 μm. The macroconidia are elliptical, 8.3–27 × 4.2–5.8 μm; the microconidia are ovoid, 4.2–6.7 × 2.5–3.3 μm. Both spore types are hyaline and nonseptate. *C. gigantea* can be isolated by surface-sterilizing mature sclerotia and plating fragments of the horny interior on glucose-yeast extract agar at 24°C. Colonies attain their maximum diameter (about 1.5 cm) in about four weeks.

Disease Cycle

The fungus overwinters as sclerotia on the ground or mixed with seed. The sclerotia germinate in the spring to produce stalked stromata with heads containing many embedded perithecia with asci (Fig. 39). The ascospores, which are hyaline and filiform, serve as primary inoculum. These spores are forcibly ejected and carried by the wind to floral parts of susceptible maize plants. The developing hyphae produce macroconidia and microconidia in a sticky matrix (honeydew). The conidia cause secondary infections. Insects may be important in conidial transmission.

Epidemiology

Ergot is endemic in certain high, cool, humid valleys of central Mexico. Disease development is favored by mean temperatures of 13–15°C and high rainfall (1,000 mm or 40 in. per year).

Selected References

FUCIKOVSKY, L., and M. MORENO. 1971. Distribution of *Claviceps gigantea* and its percent attack on two lines of corn in the state of Mexico, Mexico. Plant Dis. Rep. 55:231-233.

FUENTES, S. F., M. de LOURDES de la ISLA, A. J. ULLSTRUP, and A. E. RODRIGUEZ. 1964. *Claviceps gigantea*, a new pathogen of maize in Mexico. Phytopathology 54:379-381.

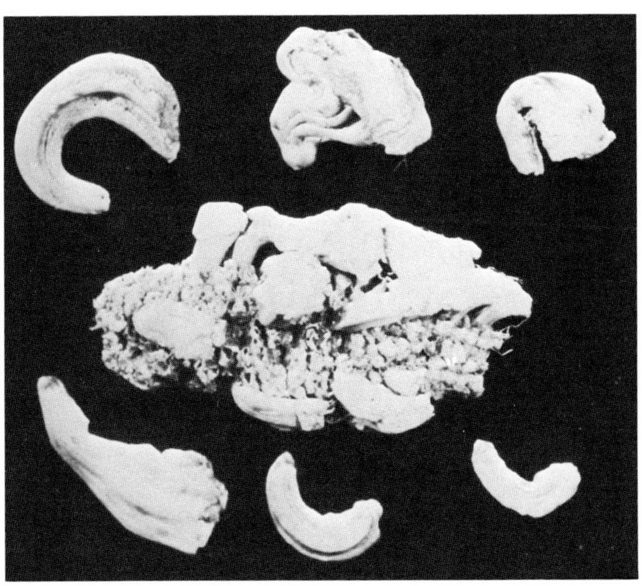

Fig. 38. Ergot or horse's tooth. (Courtesy A. J. Ullstrup)

Fig. 39. Germinating sclerotium of *Claviceps gigantea* with numerous stalked stromata. (Courtesy Carlos deLeon)

Smuts

Common Smut

Common smut or boil smut occurs worldwide wherever maize is grown. The disease is most severe in young, actively growing plants (Table II).

Losses from this disease vary greatly from one area to another and may range from a trace to (rarely) 10% or more. With resistant varieties and hybrids, annual losses seldom exceed 2% in large areas.

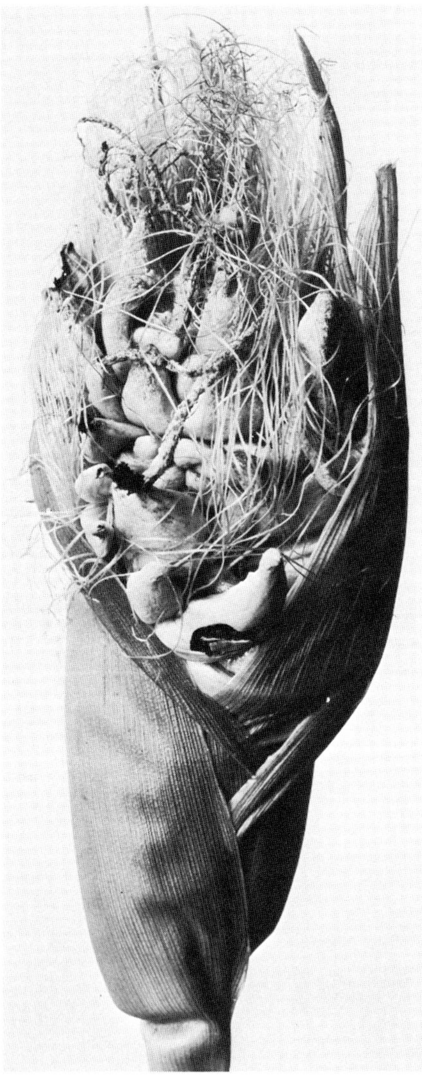

Fig. 40. Common smut. Early infection on the ear. Smut galls are still mostly covered by a silvery white membrane.

Symptoms

All aboveground parts of the plant are susceptible, particularly young, actively growing, or meristematic tissues. Symptoms are easily recognized. Galls are at first covered with a glistening, greenish-white to silvery-white tissue (Fig. 40). The interior of these galls darkens and turns into masses of powdery, dark olive-brown to black spores, except for galls on the leaves. As the galls mature, they may reach 15 cm in diameter. Galls on the leaves usually remain small, 0.6 –1.2 cm in diameter, become hard and dry but do not rupture. Early infection may kill young plants, but this is rare. Plants with galls on the lower stalks may be barren or produce several small ears. Such maize plants appear reddish in the fall, resembling those heavily infested with aphids earlier in the season.

Causal Organism

Ustilago maydis (DC.) Cda. (Syn. *Ustilago zeae* Ung.)

The teliospores (commonly called chlamydospores) are olive-brown to black, spherical to ellipsoidal, bluntly echinulate, and 8–11 μm in diameter. These diploid spores germinate by formation of a promycelium with four or more hyaline fusiform sporidia. Infection may take place from hyphae arising directly from germinating teliospores or developing from the fusion of sporidia of compatible mating types.

Disease Cycle

The disease cycle is represented in Fig. 41. The chlamydospores overwinter and germinate under favorable conditions to produce sporidia carried by air currents or splashed by water to young, developing tissues of maize. Binucleate mycelium penetrates the host through stomata, wounds, or directly through cell walls stimulating the host cells to proliferate.

Epidemiology

Development of common smut is favored by dry conditions and temperatures between 26 and 34°C. The interval between infection and gall formation varies from one to several weeks under favorable conditions. Disease incidence is higher among plants grown in soils high in N or after heavy applications of manure. Injuries due to hail, blowing soil or sand particles, cultivation, spraying, buggy-whipping, or detasseling in seed fields greatly increases the potential for smut infection. Generalized or field resistance is required because many pathogenic races or biotypes of *U. maydis* occur. All corn is susceptible to a greater or lesser degree.

Table II. Some Differences Between Common and Head Smuts

	Common Smut (*Ustilago maydis*)	Head Smut (*Sphacelotheca reiliana*)
Propagation	Windborne sporidia	Soilborne teliospores
Infection	Local; mycelium not systemic	Seedling type; mycelium systemic
Teliospores (chlamydospores)	Globose, spiny, 8–11 μm in diameter; olive-brown to black	Globose, spiny, 9–12 μm in diameter; reddish brown to black
Other characteristics	No strands in spore masses. Galls tumorlike; on any aboveground plant part	Vascular bundles of host present in sori as threadlike strands; often produces "phyllody" in floral parts

Control

1. Somewhat resistant hybrids and varieties.
2. Avoid mechanical injuries to plants during cultivation and spraying.
3. Maintain well-balanced soil fertility.
4. In home gardens, remove and burn galls from infected plants before they rupture.

Head Smut

Head smut occurs periodically in the USA in the deltas and intermountain valleys of the Pacific Coast states and the High Plains of Texas. It also is found in Australia, Mexico, New Zealand, USSR, South Africa, and southeastern Europe.

Symptoms

Head smut first appears when ears and tassels are formed (Table II). The disease is accentuated by N deficiency. It is most common in spots where N application was missed. Sporulation is most common in the ear, but frequently occurs in the tassel or on an occasional leaf. The sori are coarse and on eruption reveal dry masses of teliospores. Tassel infection may be confined to individual spikelets causing a shootlike growth, or the entire tassel may proliferate, forming bizarre structures. If the tassel is infected, all the ears on that plant will be smutted or aborted with rudiments of leafy tissue replacing the normal ears. Smutted ears are rounded and lack silks. When the tassel is *not* infected, most of the ears will be smutted, but a few nonsmutted nubbins may be produced. An occasional, partially smutted ear with silks and a few kernels may develop. Plants with smutted tassels may be severely dwarfed. Excess tillering is common. Sori on the leaves develop as long, thin strips. Infected plants do not produce pollen (Plate 31).

Causal Organism

Sphacelotheca reiliana (Kühn) Clint. (Syn. *Sorosporium reilianum* (Kühn) McAlp., *Ustilago reiliana* Kühn)

Teliospores are reddish brown to black, globose, abundantly and conspicuously spiny, and 9–12 μm in diameter. They germinate forming basidia and lateral sporidia that are small, hyaline, single-celled, and

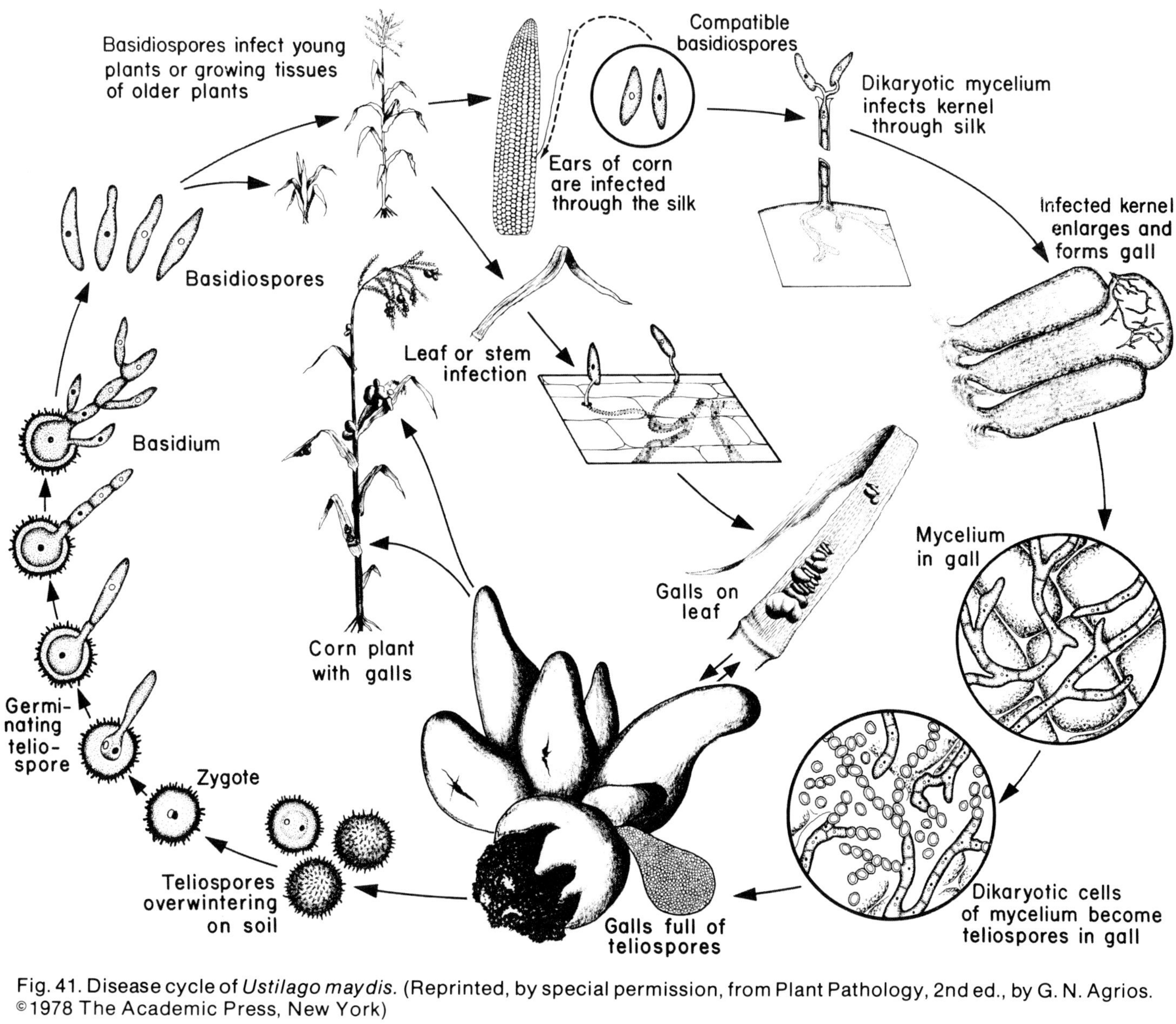

Fig. 41. Disease cycle of *Ustilago maydis.* (Reprinted, by special permission, from Plant Pathology, 2nd ed., by G. N. Agrios. ©1978 The Academic Press, New York)

subglobose, measuring 7–15 μm in diameter. Plus and minus mating types exist among the sporidia in approximately equal numbers. Infection occurs only from a sporidial mixture of both mating types. Teliospores may germinate in soil producing long, dicaryotic infection hyphae. Pathogenic specialization occurs in *S. reiliana* with one race group limited to maize and another to grain sorghums, sorgo, forage-sorghum hybrids, and some Sudangrass varieties. Experiments have identified a hybrid of *S. reiliana* that infects both maize and sorghum.

Disease Cycle

S. reiliana is primarily soilborne. Teliospores from smutted heads overseason in soil and are the main source of infection. The fungus attacks seedlings and the mycelium develops systemically, invading undifferentiated floral tissues. Part or all of these tissues develop into smut sori. The sori are covered with a thin tissue that breaks open to expose the spore masses. Fine, threadlike strands—vestiges of the vascular system of the maize inflorescence—ramify through the sori and are characteristic of infection by *S. reiliana*. Secondary effects of infection include dwarfing, sterility, and phyllody. Although the teliospores may be seedborne, this is not an important source of inoculum.

Epidemiology

The infection level is related to the concentration of teliospores in the soil. Spores survive adverse conditions for long periods. Soil temperatures of 21–28°C and moderate to low soil moisture (15–25% w/w) are optimal for seedling infection. Other hosts of *S. reiliana* are pitscalegrass, sorghum, and Sudangrass.

Fig. 42. False smut. (Courtesy Carlos deLeon)

Control

1. Resistant hybrids and varieties. Most US hybrids are highly tolerant.
2. Chemical treatment of soil in rows at or before planting time may be feasible in seed-production fields.
3. Sanitation and crop rotation.
4. Seed treatment.

False Smut

False smut has been reported worldwide. It occurs on rice, maize, and other Gramineae but is of little economic importance.

Symptoms

False smut is characterized by galls (sclerotia) similar to ergot that replace the flowers on the tassels. Only a few flowers are usually infected. The sclerotia superficially resemble smut sori (Fig. 42).

Causal Organism

Ustilaginoidea virens (Cke.) Tak. (Syn. *U. oryzae* (Pat.) Bref.)

Mature sclerotia are spherical, 4–15 mm in diameter, olive-green to black, and velvety with white interiors. The conidia are spherical to oval, echinulate, olive-green, 4–7 μm in diameter, and borne on short, sterigmalike projections from septate, yellowish-green hyphae.

Disease Cycle

Similar to ergot (page 37).

Epidemiology

False smut is favored by hot, wet weather.

Selected References

CHRISTENSEN, J. J. 1963. Corn smut caused by *Ustilago maydis*. Monograph No. 2. Am. Phytopathol. Soc.: St. Paul, MN.

FISCHER, G. W. 1953. Manual of the North American Smut Fungi. Ronald Press: New York.

FISCHER, G. W., and C. S. HOLTON. 1957. Biology and Control of the Smut Fungi. Ronald Press: New York.

FREDERIKSEN, R. A. 1977. Head smut of corn and sorghum, pp. 89-105. Proc. 32nd Annu. Corn and Sorghum Res. Conf., Am. Seed Trade Assoc.: Washington, DC.

HALISKY, P. M. 1963. Head smut of sorghum, sudangrass, and corn, caused by *Sphacelotheca reiliana* (Kühn) Clint. Hilgardia 34:287-304.

HASKELL, R. J., and W. W. DIEHL. 1929. False smut of maize, *Ustilaginoidea*. Phytopathology 19:589-592.

Rusts

Common Maize Rust

Symptoms

Pustules (sori) may appear on any aboveground part, being most abundant on the leaves. The pustules occur nearly simultaneously on both leaf surfaces in contrast with southern rust (see below), which has very sparse pustule development on the lower leaf surface. The circular to elongate, golden-brown to cinnamon-brown pustules are sparsely scattered over both leaf surfaces becoming brownish black as the plant matures and teliospores develop. When severe, chlorosis and death of the leaves and leaf sheaths may occur. The pustules become erumpent and powdery early in their development (Plate 32).

Causal Organism

Puccinia sorghi Schw.

The uredospores are cinnamon brown, mostly spherical to broadly ellipsoidal, and 21–30 × 24–33 μm. The spore walls are golden or cinnamon brown, 1.5–2 μm thick, and moderately echinulate with three to four equatorial germ pores. Each spore is binucleate as is the mycelium that develops on germination. The teliospores, which replace the uredospores in the pustules, are chestnut brown to golden brown, smooth, oblong to ellipsoid or obovoid, two-celled with slight constrictions at the septa, measure 14–25 × 28–46 μm, and are attached to pale yellowish to brownish pedicels (up to 80 μm long) once or twice the length of the spores (Fig. 43A). The aeciospores are pale yellow, verrucose, mostly spherical or ellipsoid, 13–19 × 18–26 μm, and occur in "cluster-cups" on species of *Oxalis*. *P. sorghi* can be cultured on detached maize leaves floated on a solution containing 5% sucrose and 20 ppm of kinetin.

Disease Cycle

Teliospores germinate in the spring in certain areas of the world to form basidia on which small, thin-walled, hyaline, haploid basidiospores are produced. These spores germinate and penetrate the leaves of wood sorrel (*Oxalis* spp.) forming spermagonia with minute spermatia on the upper leaf surface. The spermatia fuse with flexuous (receptive) hyphae of the opposite mating type initiating the aecial stage on the lower surface of *Oxalis* leaves. The binucleate aeciospores in the "cluster-cups" are windborne and infect maize leaves. These infections give rise to uredospores, the repeating stage of the fungus. In most temperate areas of the world, including the USA, the fungus does not infect *Oxalis*. Spores are windblown into temperate regions from nearby subtropical and tropical areas where the pathogen persists on living maize plants.

Epidemiology

Cool temperatures (16–23° C) and high relative humidity (100%) favor rust development and spread. Some inbred lines show a resistant flecking reaction to infection. Older maize tissue is generally resistant to *P. sorghi*. Numerous physiologic races of *P. sorghi* can be separated by their reactions on lines of maize having specific genes for resistance. Resistance in mature plants is polygenically controlled.

Geographical Distribution and Hosts

Aecial infection occurs infrequently in temperate regions of Europe, USSR, USA, Mexico, South Africa, India, and Nepal on *Oxalis* spp. Uredial and telial stages occur worldwide where maize or teosinte is grown.

Control

1. Resistant hybrids and varieties.
2. Fungicide applications, starting when pustules first appear on the leaves, may be feasible, especially in seed-production fields.

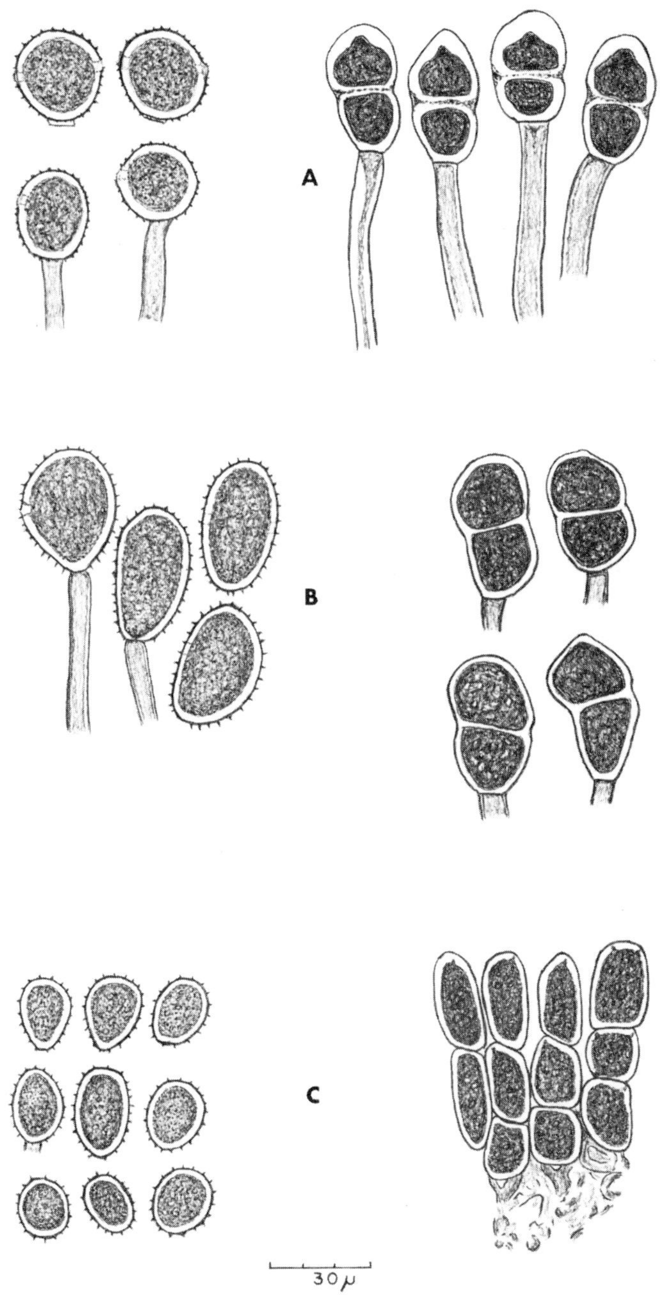

Fig. 43. Uredospores and teliospores of three species of rust fungi: **A,** *Puccinia sorghi;* **B,** *P. polysora;* **C,** *Physopella zeae* (after Lentz).

Southern (Polysora) Rust

Symptoms

This disease resembles common rust but with very subtle differences. Pustules are light cinnamon brown, circular to oval, 0.2–2.0 mm long, and densely scattered on the upper leaf surface. In nature, pustule development on lower leaf surfaces occurs more slowly and less abundantly than on upper leaf surfaces. On maize grown in the glasshouse, pustule development rarely occurs on lower leaf surfaces. On maize grown in growth chambers, southern rust infections are very rare on plants more than a month old. The telia are circular to elongate, 0.2–0.5 mm in diameter, chocolate brown to black, and remain covered by the epidermis longer than in common rust. The telia often appear in circles around the uredial pustules. Infected maize leaves become chlorotic and dry (Plate 33).

Causal Organism

Puccinia polysora Underw.

The uredospores are yellowish to golden, mostly ellipsoid or obovoid, and measure 20–29 × 29–40 μm. The golden or yellowish walls are minutely and sparsely echinulate, 1–1.5 μm thick, with four or five equatorial pores. The teliospores are chestnut brown, smooth, angularly ellipsoid or oblong, rounded at both ends, and 18–27 × 29–41 μm. They are very brittle, usually two-celled, constricted at the septum, and borne on short (10–30 μm), persistent, yellow or brownish pedicels one-fourth or less the length of the spore (Fig. 43B).

Disease Cycle

Teliospores of *P. polysora* are rare and are not known to germinate. They appear to be unimportant in the disease cycle. The uredospores constitute both primary and secondary inoculum, and are carried to maize plants by wind or on infected materials. No alternate host for *P. polysora* has been found. Thus, only uredial and telial stages of this rust are known.

Epidemiology

Southern rust is favored by high temperatures (27°C) and high relative humidity. Elevations above 4,000 ft (1,220 m) are unfavorable for disease development while those below 3,000 ft (900 m) favor early rust onset. Several races of *P. polysora* have been reported. Resistance to each race is governed by monogenic dominants.

Geographical Distribution and Hosts

Southern rust occurs in East, West, and South Africa, Southeast Asia, The Philippines, Australia, southern and central North America, Central and South America, and the West Indies. In the USA, southern rust occurs principally in the southeastern states but has been reported as far north as Indiana, Illinois, and southern Wisconsin. Hosts besides maize include plumegrasses, teosinte, and gamagrasses.

Tropical Rust

Symptoms

The pustules are yellow, occur in small groups mostly 0.3–1.0 mm long on the upper leaf surfaces, and are *covered* by the epidermis except for a small pore or slit. The pustules later develop into dark purplish, circular or oblong blotches with creamy centers, 0.6 cm in diameter (Plate 34). The dark, chocolate-brown to blackish telia are usually scattered in groups around the uredia.

Causal Organism

Physopella zeae (Mains) Cummins & Ramachar. (Syn. *Angiopsora zeae* Mains)

The uredospores are sessile, ellipsoid or obovoid, 15–22 × 22–33 μm, colorless to pale yellow, with nearly hyaline, moderately echinulate walls, 1.5–2 μm thick. The pores are very obscure, probably seven or eight, and scattered. The teliospores are cylindrical, golden brown to chestnut brown, one-celled, occur in chains of two to three, and measure 12–18 × 16–36 μm. The oblong to angularly ellipsoid teliospores are sessile with smooth, golden-brown walls 1.5–2 μm thick at the sides and 2.5–6 μm at the apex of apical spores.

Disease Cycle

See southern rust (above). Aecia are unknown. No alternate host has been reported.

Epidemiology

The disease is favored by warm to hot, humid weather and low altitudes. At least two races of *P. zeae* are known.

Geographical Distribution and Hosts

Tropical rust appears to be limited to warm, humid areas of Mexico, Central and northern South America, and the Caribbean. It has not been reported in the USA. Teosinte is another host.

Control

Resistant hybrids and varieties.

Identifying Rusts

Puccinia sorghi can be differentiated from *P. polysora* by the larger, more sparsely distributed and elongated uredia; darker uredospores; erumpent telia; and teliospores with thicker walls, longer pedicels, and a bullet-shaped terminal cell. The uredia of *P. sorghi* are common on both leaf surfaces; those of *P. polysora* are much more numerous on the upper leaf surface. The telia of *P. polysora* remain covered. *Physopella zeae* is identified by hyaline uredospores; covered, blackish telia; and sessile teliospores in short chains (Fig. 43C).

Selected References

CAMMACK, R. H. 1959. Studies on *Puccinia polysora* Underw. III. Description and life cycle of *P. polysora* in West Africa. Br. Mycol. Soc. Trans. 42:55-58.

CUMMINS, G. B. 1941. Identity and distribution of three rusts of corn. Phytopathology 31:856-857.

CUMMINS, G. B. 1971. Rusts of cereals, grasses and bamboos. Springer-Verlag: New York.

HOOKER, A. L., and C. E. YARWOOD. 1966. Culture of *Puccinia sorghi* on detached leaves of corn and *Oxalis corniculata*. Phytopathology 56:536-539.

ROBERT, A. L. 1962. Host ranges and races of the corn rusts. Phytopathology 52:1010-1012.

SCHIEBER, E. 1971. Distribution of *Puccinia polysora* and *P. sorghi* in Africa and their pathogenicity on Latin American maize germ plasm. FAO Plant Prot. Bull. 19:25-31.

SCHIEBER, E., and J. G. DICKSON. 1963. Comparative pathology of three tropical corn rusts. Phytopathology 53:517-521.

Stalk Rots

Stalk rots are universally important and among the most destructive diseases of maize throughout the world (Table III). Development of stalk rots is favored by an early environment that encourages kernel set and a late environment that is stressful. Postflowering stresses include various leaf blights, excessive cloudiness, crowded plants, drought, hail damage, low K coupled with high N, and injury by stalk-boring and root-boring insects, all of which lower the amount of photosynthate produced by the plant.

In most cases, stalk rots are caused by a complex of several species of fungi and bacteria that attack plants approaching maturity. *Pythium* and bacteria (Table III) commonly attack maize before anthesis. Identification of a specific stalk rot is therefore difficult.

Losses vary from season to season and from region to region. Yield losses of 10 to 20% may occur on susceptible hybrids in the USA, while losses of 25 to 33% occur in other countries where certain pathogens are particularly destructive. Tolls greater than 50% have been reported in localized areas. Losses may be direct (due to poor filling of ears or lightweight and poorly finished ears) or indirect through harvest losses because of stalk breakage or lodging (Fig. 44).

Colletotrichum or Anthracnose Top-Dieback and Stalk Rot

Since the early 1970s, corn anthracnose has become much more prevalent in the USA, especially in the southeastern states and the southern half of the Corn Belt. Leaf blight (page 22) and stalk rot are the most common phases of the disease. However, the causal fungus can infect most plant parts including the shanks, kernels, and roots.

Symptoms

Stalk infections become evident at various stages of growth depending on the susceptibility of the inbred, hybrid, or variety. The lower stalk tissues of highly susceptible plants may be so severely rotted that the plants are killed before pollination. Most hybrids and inbreds, however, are not affected until shortly before normal senescence.

Occasionally, portions of the plant above the ear blanch and die four to six weeks after pollination while the lower portions of the plant remain green. Top dieback symptoms are generally expressed late in the season sometime after the dough stage (Plate 35). The upper two or three leaves may turn yellow or reddish, lodge, and drop off. The upper leaf discoloration often occurs at about the time the lower leaves begin to senesce. In other cases, entire plants may die prematurely and later lodge. Stalk death and lodging are often expressed later than top dieback. Lodging is normally found higher on the stalk than with other stalk rot diseases.

Stalk symptoms generally appear after tasseling as narrow, vertical or oval, water-soaked lesions in the rind. These lesions become tan to reddish brown and finally dark brown to black late in the season, and may cover the lower internodes or the entire stalk. Black specks (acervuli), associated with the production of spores by the causal fungus are often found on the stalk rind. The stalk lesions coalesce forming relatively large, dark brown to shiny black, blotchy areas or streaks that may be somewhat sunken (Fig. 45). The external lesions are accompanied by a brown or black discoloration of the pith tissue starting at the nodes. The pith soon becomes disintegrated (Fig. 46). With some genotypes the pith may be soft, watery, and easily crushed.

Causal Organism

Colletotrichum graminicola (Ces.) G. W. Wils. See page 23 for a full description of the fungus.

Disease Cycle

The causal fungus overseasons as conidia in acervuli on maize leaf and stalk debris and as dormant resting mycelial cells in stalk debris. The conidia are the primary inoculum for the leaf blight phase (page 22). Primary inoculum for the stalk rot phase is not completely understood. Some researchers believe that conidia washed behind the leaf sheaths germinate and invade the stalk. Stalk rot occurs in fields of maize with very little anthracnose leaf blight and with little or no debris on the soil surface, however. Indications are that

Fig. 44. Stalk rot. Severe stalk breakage caused by stalk-rotting fungi. (Courtesy G. L. Worf)

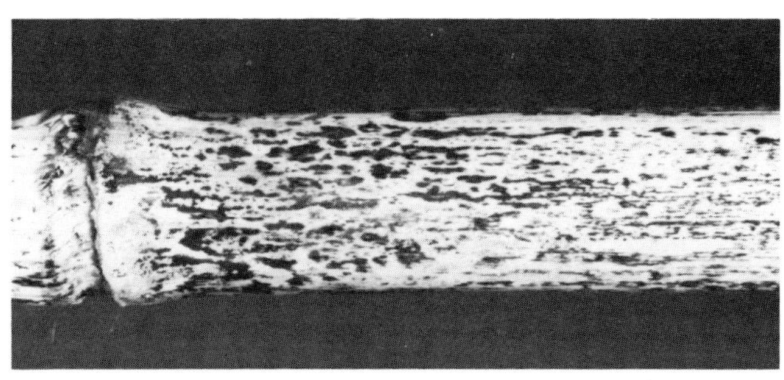

Fig. 45. Anthracnose. Stalk lesions become shiny black, blotchy areas or streaks. (Courtesy H. L. Warren)

inoculum for stalk rot may come partially from buried stalk debris where resting cells can survive for several years. The fungus invades maize roots, growing into the old stalk debris as well as insect feeding sites. The fungus may then colonize the lower internodes as normal senescence progresses.

Epidemiology

See page 23. Disease severity is influenced by the susceptibility of the hybrid and high temperatures.

Control

1. Resistant hybrids and varieties. Resistance to stalk rot caused by *C. graminicola* is not highly correlated with resistance to other stalk rot fungi.
2. Crop rotation, clean and deep plowdown of crop residues, and careful use of minimum or conservation tillage.
3. Balanced soil fertility.

Diplodia Stalk Rot

Symptoms

The disease commonly appears several weeks after silking. Affected plants die suddenly, leaves wilt, become dry, and appear grayish-green resembling frost injury. The lower internodes are brown to straw-colored, spongy, and easily crushed. The pith disintegrates and becomes discolored with only the vascular bundles remaining intact (Fig. 47). A characteristic sign is the presence of *subepidermal,* minute, dark brown to black pycnidia clustered near the nodes. White fungal growth may also be present on the surface.

Causal Organism

Diplodia maydis (Berk.) Sacc. (Syn. *D. zeae* (Schw.) Lev.)

The fungus produces globose or flask-shaped pycnidia containing olive-colored to brown, elliptical, two-celled, straight to slightly curved spores, 5–6 × 25–30 μm in size (Fig. 48A). Less commonly, pycnidia contain colorless, long, narrow, threadlike spores, 1–2 × 25–35 μm. No sexual stage of the fungus is known. *D. macrospora* Earle and *D. frumenti* Ell. & Ev. also occur on maize (Fig. 48B and C). *D. zeicola* Saccas attacks maize in Africa.

Disease Cycle

D. maydis overseasons as spores in pycnidia on debris and as spores or mycelium on the seed. Under warm, moist conditions, spores are extruded from pycnidia in long cirri and disseminated by rain, wind, and probably by insects. Infection of maize plants occurs primarily through the crown, mesocotyl, and roots or occasionally at the nodes between the crown and ear. The pathogen then grows into the stalks. Seedborne inoculum may cause seedling blight. The fungus does not invade the entire plant.

Epidemiology

Dry conditions early in the season and warm (28–30°C), wet weather two to three weeks after silking favor development of Diplodia stalk rot. High N and low

Table III. Fungal and Bacterial Stalk Rots of Maize

Common Name	Pathogen	Host Range	Geographic Distribution	Characteristic Symptoms
Colletotrichum or anthracnose top-dieback and stalk rot	*Colletotrichum graminicola*	Maize and other grasses	USA, South Africa, Europe, South and Central America	Acervuli on stalks; narrow, water-soaked lesions followed by black streaks or blotches
Diplodia stalk rot	*Diplodia maydis, D. macrospora*	Maize	USA, West and South Africa, Australia, Philippines, Romania	Pycnidia embedded in stalks; pith shredding; dry basal stalk rot
Gibberella stalk rot	*Gibberella zeae (Fusarium roseum* f. sp. *cerealis)*	Maize, wheat, oats, rice, barley, and others	Worldwide	Superficial perithecia on stalks; internal reddish stalk discoloration
Fusarium stalk rot	*F. moniliforme (G. moniliformis), F. moniliforme* var. *subglutinans*	Maize and others	Worldwide	Salmon-colored mycelium on stalk surface under warm, moist conditions; pith disintegration starting at the nodes
Charcoal rot	*Macrophomina phaseolina (Sclerotium bataticola)*	Maize, sorghums, cotton, soybeans, and others	North America, Europe, South Africa	Numerous small, black sclerotia on vascular strands and roots
Pythium stalk rot	*Pythium aphanidermatum*	Many hosts	Worldwide	Stalks collapse; twisted, water-soaked; brown root rot
Late wilt	*Cephalosporium maydis*	Maize, cotton, lupine	Egypt, India	Rapid wilting of leaves; vascular bundles discolored; stalk dry, shrunken, and hollow
Black bundle	*Cephalosporium acremonium*	Maize, sorghum, soybean, wheat, oats, and others	USA, Egypt, India	Blackening of vascular bundles; purpling of leaves and stalks
Bacterial stalk rot	*Erwinia chrysanthemi* pathovar *zeae*	Maize and others	Worldwide	Stalks collapse; twisted, disintegrated into soft mass; often with a foul odor

K, high plant populations, and loss of leaf area through disease, hail, or insect damage predispose plants to infection. Early-maturing hybrids are generally more susceptible than full-season hybrids.

Control

1. Resistant hybrids and varieties.
2. Balanced soil fertility; avoid high levels of N and low levels of K.
3. Lower plant populations.

Gibberella Stalk Rot

Symptoms

Leaves on early infected plants suddenly turn a dull grayish-green while the lower internodes soften and turn tan to dark brown. The stalks often show an internal pink to reddish discoloration of the diseased tissue (Plate 36). The fungus causes shredding of the pith (Fig. 49) and may produce small, round, black perithecia superfically on the stalks (Fig. 50 and Plate 36). Lesions may develop concentric rings. This disease can be differentiated from Diplodia stalk rot by pith discoloration and the *superficial* perithecia.

Causal Organism

Gibberella zeae (Schw.) Petch. (Asexual state, *Fusarium roseum* f. sp. *cerealis* 'Graminearum'); (Syn. *F. graminearum* Schwabe)

The perithecia are bluish black, spherical, and borne superficially on diseased stalks. When fruiting structures mature, asci develop, each containing eight ascospores arranged obliquely in one row. The ascospores are hyaline, three-septate, tapering uniformly to the ends, slightly curved, and 3–5 × 20–30 μm (Fig. 51A).

The macroconidia of *F. roseum* 'Graminearum' are hyaline, curved, and tapering toward the tips, three- to five-septate, and 4–6 × 30–60 μm (Fig. 51B). The mycelium appears pinkish white on PDA and produces no microconidia. Some isolates produce chlamydospores.

Fig. 46. Anthracnose stalk rot caused by *Colletotrichum graminicola;* healthy plant at left. (Courtesy D. G. White)

Disease Cycle

Perithecia on infected maize stalks mature under warm, wet conditions. Mature ascospores are exuded and carried by air currents to ears or stalks where they germinate and penetrate the tissues. Conidia are produced on the pinkish-white mycelium that grows on diseased plant parts during warm, moist weather. The fungus may overseason in maize debris and rarely in seed. The causal fungus also infects wheat, barley, and other cereals where it causes diseases known as scab and seedling blight.

Epidemiology and Control

Same as for Diplodia stalk rot (above).

Fusarium Stalk Rot

This rot is difficult to distinguish from Gibberella stalk rot. Rotting commonly affects the roots, plant base, and lower internodes. Rot normally begins soon after pollination and becomes more severe as the plant matures. A whitish-pink to salmon discoloration of the pith, stalk breakage, and premature ripening are the same as for Gibberella stalk rot (Plate 37).

Causal Organisms

Fusarium moniliforme Sheld. (Perfect state *Gibberella moniliformis* (Wineland); Syn. *Gibberella fujikuroi* (Saw.) Wr. and *F. moniliforme* var. *subglutinans* Wr. &

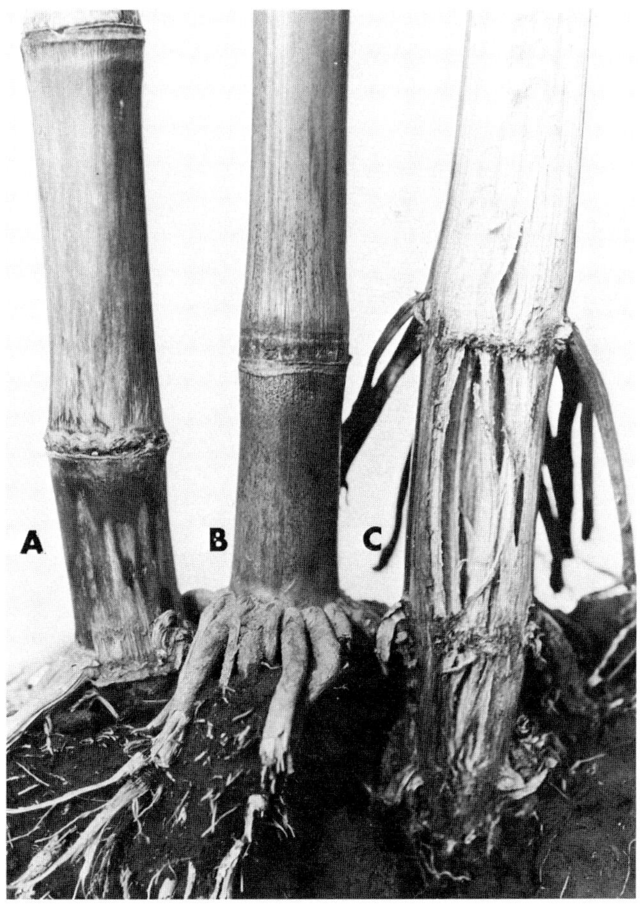

Fig. 47. Diplodia stalk rot. **A,** External discoloration; **B,** pycnidia of *Diplodia maydis* clustered just below the node; **C,** stalk split to show disintegration of tissues near soil level.

Reink.)

Both fungi produce asexual spores on mycelium, which appears as a cottony-pink growth on the leaf sheaths and at the nodes. The macroconidia of *F. moniliforme* are sparse, three- to seven-septate, curved toward the tips, and 2.4–4.9 × 15–60 μm (Fig. 51D). Microconidia, 2–3 × 5–12 μm in size, are produced in abundance in chains (Fig. 52) or false heads on hyphal branches.

The perithecia, rarely observed in the USA, are globose, smooth, and blue-black. The asci are oblong, 75–100 × 10–16 μm, and contain eight ascospores that are straight, tapering to tips, constricted at the septa, mostly one-septate, 4.5–7.0 × 12–17 μm, and arranged in two irregular rows.

The macroconidia of *F. moniliforme* var. *subglutinans* are slightly less curved than those of *F. moniliforme* and usually are three-septate. The microconidia are borne singly or in false heads, never in chains. The rarely found perfect stage produces asci commonly containing eight ascospores (sometimes four or six). They are rounded at the tips, usually one septate, rarely two- to three-septate, 4.5–7.0 × 12.1–17 μm, and arranged in one oblique row (Fig. 51C).

Disease Cycle

The fungi develop on crop residue in or on the soil surface. Under favorable conditions, fungi infect maize stalks either directly or through wounds caused by hail or insects. *F. moniliforme* can penetrate stalks at the base of the leaf sheaths and progress into the lower internodes. The fungus is commonly seedborne, but this source of inoculum is believed to be less important than airborne or soilborne inoculum. Ears, kernels, roots, or seedlings may also be infected.

Epidemiology and Control

Same as for Diplodia stalk rot (above).

Pythium Stalk Rot

This rapidly developing rot is usually confined to a single internode just above the soil line. The stalks are *not* completely broken off and the diseased area is brown, water-soaked, soft, and collapsed. The stalks may also be twisted and distorted. Infected plants remain green and turgid up to several weeks because the vascular bundles remain intact (Plate 38). Pythium stalk rot usually becomes evident shortly before or after maize has tasseled, although young, vigorous plants can be attacked.

Causal Organism

Pythium aphanidermatum (Eds.) Fitzp. (Syn. *P. butleri* Subr.) and other *Pythium* species

P. aphanidermatum can be cultured on water or

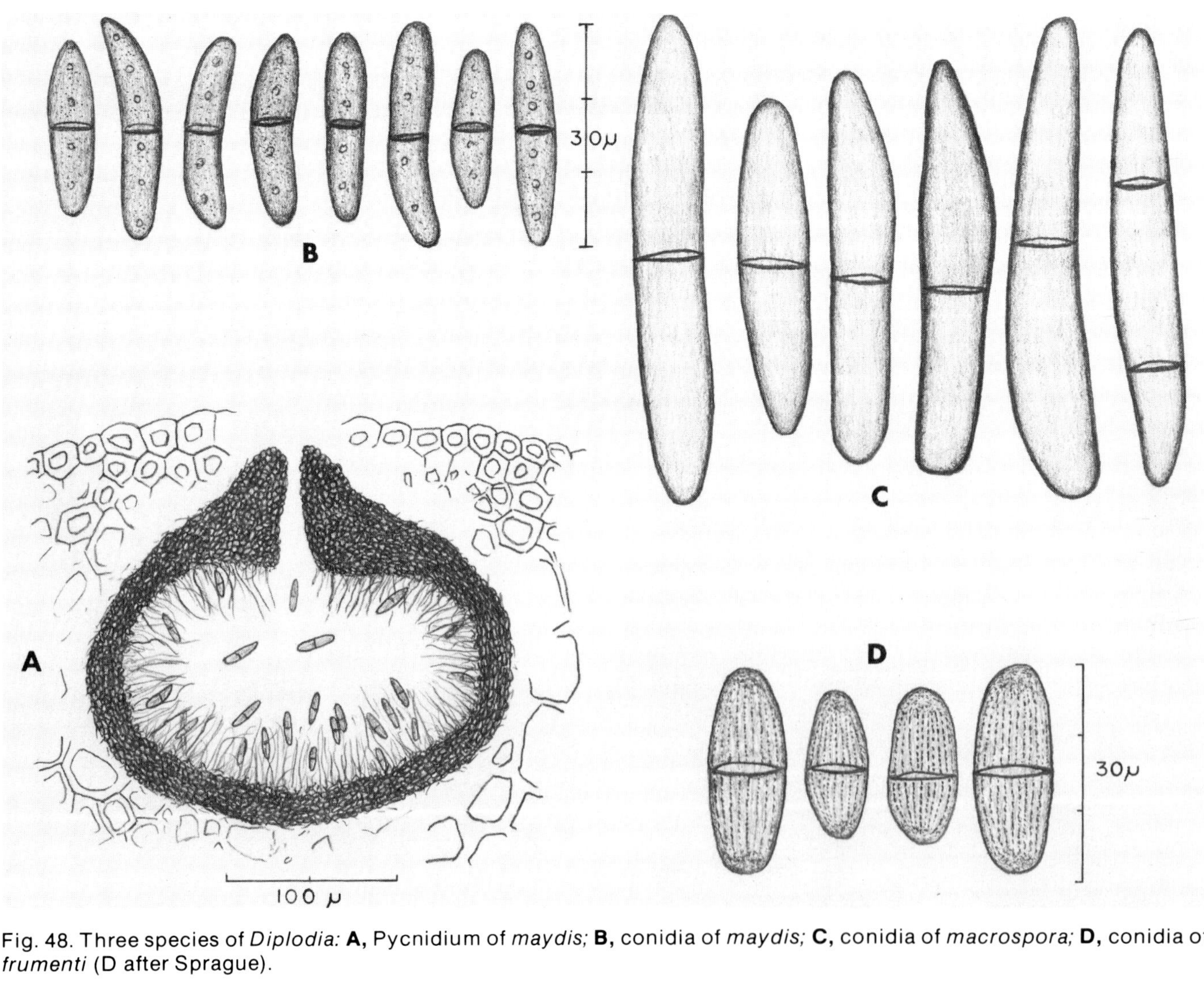

Fig. 48. Three species of *Diplodia:* **A,** Pycnidium of *maydis;* **B,** conidia of *maydis;* **C,** conidia of *macrospora;* **D,** conidia of *frumenti* (D after Sprague).

cornmeal agar. The hyphae are hyaline and nonseptate, except near fruiting structures. Sporangiophores are filamentous, branched or unbranched, measure 50–1,000 × 4–20 μm, and give rise to sporangia that bear reniform, biciliate zoospores 12 × 7.5 μm in size. Oogonia are terminal, spherical, and 22–27 μm in diameter. Antheridia are terminal or intercalary, barrel-shaped, and 9–11 × 10–14 μm in size. Oospores are 17–19 μm in diameter and thick-walled.

Disease Cycle

See seed rots and seedling blights (page 13).

Epidemiology

The disease develops primarily during extended periods of hot (32° C), wet, or very humid weather. It is frequently encountered in fields where air and soil drainage are poor and humidity is high. Unlike other stalk-rotting fungi, *Pythium* can attack plants before anthesis.

Fig. 49. Gibberella stalk rot. The pith, which becomes progressively more shredded, often shows pink to reddish discoloration.

Fig. 50. Close-up of the small, round, black perithecia of *Gibberella roseum* f. sp. *cerealis*. Perithecia develop superficially on the stalk.

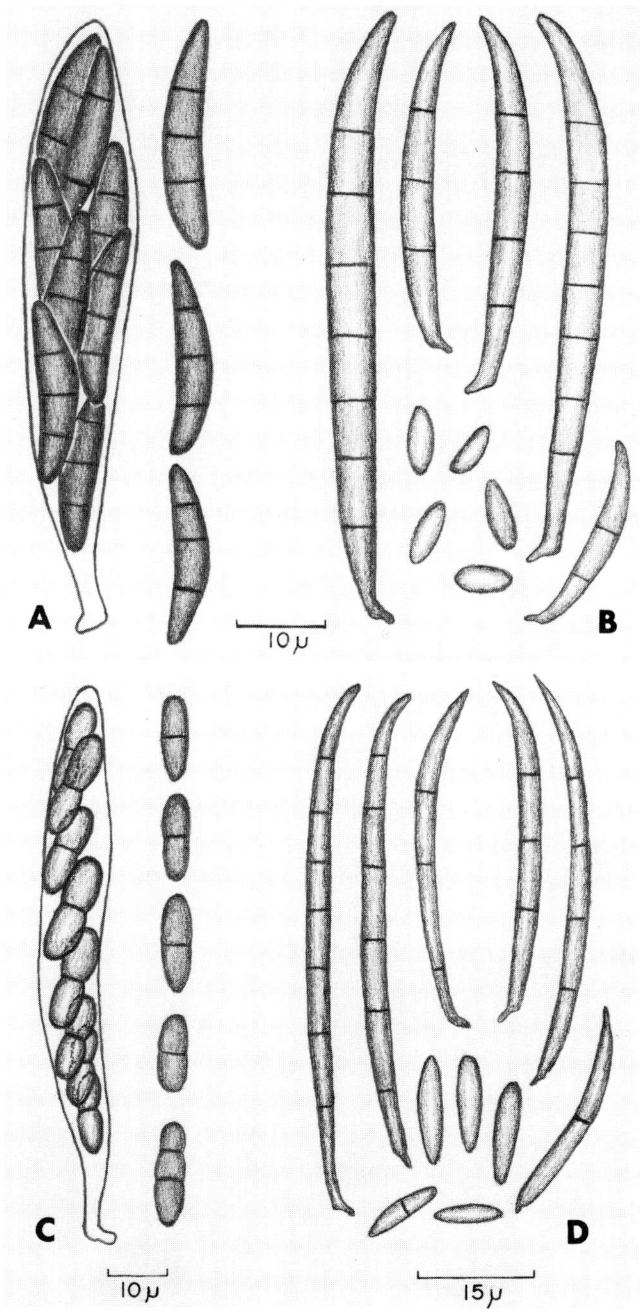

Fig. 51. **A,** Ascus and ascospores of *Gibberella zeae;* **B,** macroconidia and microconidia of *Fusarium roseum* f. sp. *cerealis* 'Graminearum'; **C,** ascus and ascospores of *Gibberella moniliformis;* **D,** macroconidia and microconidia of *Fusarium moniliforme* (after Saccas).

Control

Resistant hybrids and varieties.

Late Wilt

Symptoms

The first symptom of late wilt, which has only been reported from Egypt and India, is moderately rapid wilting of the leaves beginning at tasseling time. The leaves turn dull green and then dry. Vascular bundles in the stalk are discolored. Later, lower portions of the stalk become dry, shrunken, and hollow. Stalk symptoms may be modified by secondary organisms that cause wet rot (Plate 39).

Causal Organism

Cephalosporium maydis Samra, Sabet & Hingorani

The conidiophores are 30–250 μm long, mostly branched, with conidia forming successively and exogenously in the heads. The conidia are hyaline, single-celled, oblong, and measure 3.6–14 × 3–3.6 μm (av. 7.2 × 3.5 μm) (Fig. 53A). They germinate by one to three polar germ tubes that frequently anastomose. On PDA and at 22–24° C, colonies are white to pale gray, turning slate gray to black with age. The fungus can also be grown on oatmeal, malt extract, and yeast extract-glucose agars. Small, sclerotiumlike bodies appear in old cultures.

Disease Cycle

C. maydis is primarily soilborne and may infect maize through the roots or mesocotyl. The fungus produces appressoriumlike structures at infection sites causing infected tissues to collapse. After penetration, *C. maydis* spreads toward and develops in the xylem. Growth is slow at first but accelerates when tassels emerge. The fungus is also seedborne and may cause seed rot and preemergence damping-off.

Epidemiology

C. maydis infects maize seedlings more easily than older plants. The disease develops more rapidly in light, sandy, and heavy clay soils than in light loam and loam soils. Moist soils, control of stalk borers, and K fertilization reduce infection by the late wilt fungus.

Fig. 52. Chains of microconidia of *Fusarium moniliforme*. (Courtesy H. L. Warren)

Control

1. Resistant hybrids and varieties.
2. Rotation with rice or sesame.
3. Balanced soil fertility based on a soil test.

Charcoal Rot

Symptoms

Charcoal rot commonly attacks seedlings or plants approaching maturity. The fungus produces brown, water-soaked lesions on the roots that later turn black. As the plant matures, the fungus spreads into the lower internodes of the stalk, causing premature ripening, shredding, and breaking at the crown. Numerous black sclerotia on the vascular strands give the interior of the stalks a charred appearance and are a good diagnostic sign (Fig. 54A). Sclerotia may be found just beneath the stalk surface (Fig. 54B) and on the roots. The fungus also infects the kernels, causing them to blacken completely.

Causal Organism

Macrophomina phaseolina (Tassi) G. Goid. (Syn. *Macrophomina phaseoli* (Maubl.) Ashby and *Botryodiplodia phaseoli* (Maub.) Thirum.) (*Sclerotium bataticola* Taub.)

The minute sclerotia are black, numerous on diseased plant parts, and globular to irregular in shape (Fig. 53B). The fungus is composed of many strains, differentiated by sclerotial size and the presence or absence of pycnidia. The strains attacking maize apparently do not produce spores although other strains produce hyaline, nonseptate conidia in pycnidia. *S. bataticola,* the sclerotial and mycelial stage of *M. phaseolina,* is considered to be identical with *Rhizoctonia bataticola.*

Disease Cycle

M. phaseolina overwinters and is disseminated as sclerotia. The fungus enters through the roots and grows through the cortical tissues into the stalks as the plants mature. Seedling blight may also occur.

Epidemiology

Soil temperatures near 37° C are favorable for disease development, while either low soil temperatures or high soil moisture decreases severity. The organism has a wide host range including sorghums and soybeans.

Control

1. In irrigated areas, charcoal rot can be minimized by maintaining moist soils during dry periods after tasseling.
2. For other measures, see Diplodia stalk rot (above).

Black Bundle Disease

Symptoms

Infection becomes apparent when maize has reached the dough stage. One of the first symptoms is a purpling or reddening of the leaves and stalks. The most characteristic symptom is the blackening of the vascular bundles in the stalk extending through several internodes. Barren plants, excessive tillering, nubbin ears, and multiple ears at one node are other symptoms.

Causal Organism

Cephalosporium acremonium Cda.

The spores are nearly hyaline, one-celled, elliptical or oblong, straight or curved, 3.2–9.6 × 1.5–3.2 μm, and borne in groups in a slimy matrix at the apex of short conidiophores (Fig. 53A). The spores germinate by one

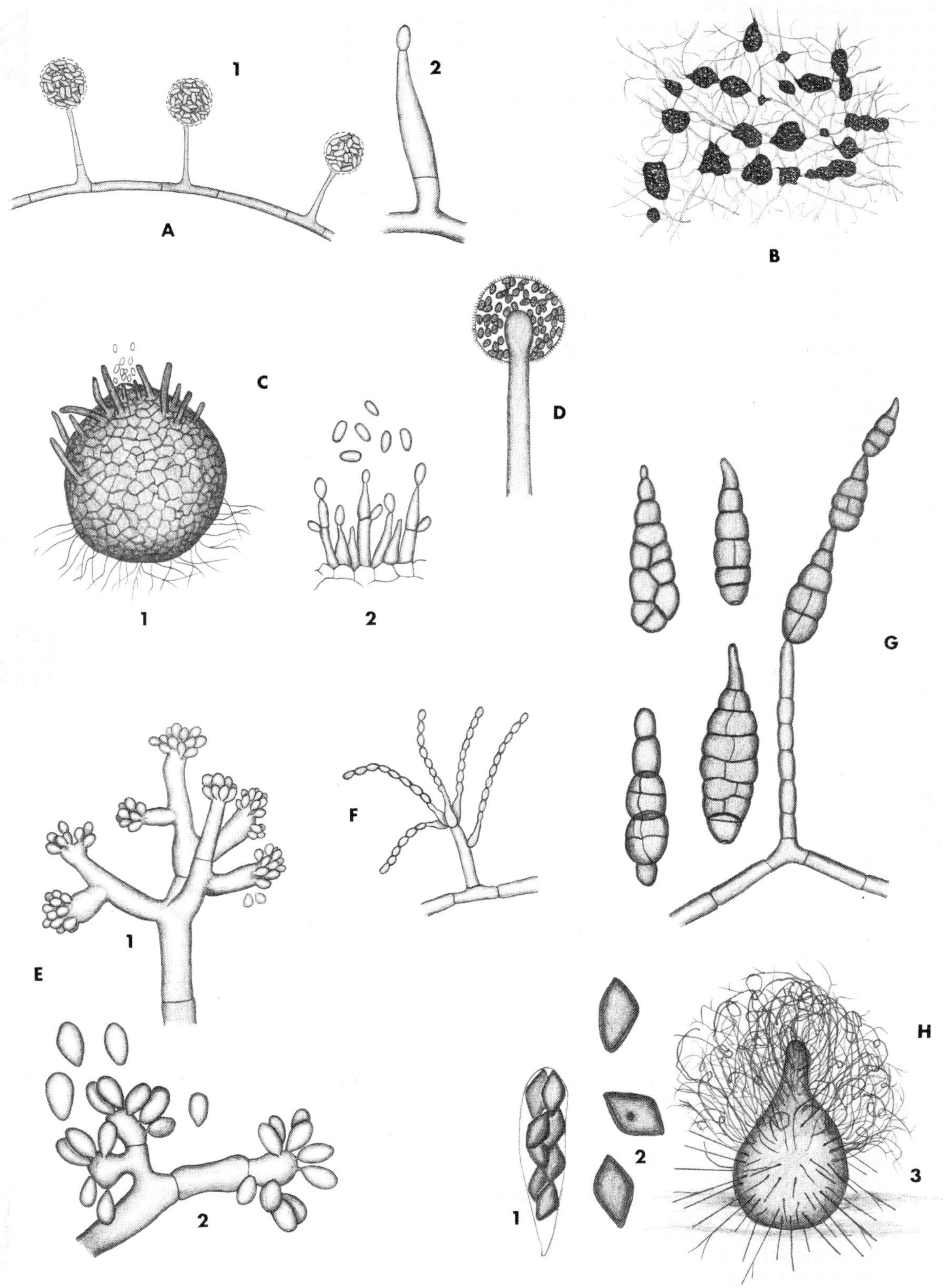

Fig. 53. Stalk-invading fungi. **A,** *Cephalosporium:* **1,** conidia in slime droplets; **2,** detail of spore production. **B,** *Macrophomina phaseoli* (*Sclerotium bataticola*): Mass of sclerotia and mycelium. **C,** *Pyrenochaeta:* **1,** pycnidium, **2,** conidiophores and conidia or pycnidiospores. **D,** *Mucor:* Sporangium with columella and sporangiospores. **E,** *Botrytis:* **1,** conidial head, **2,** conidiophores and conidia more highly magnified. **F,** *Spicaria:* Conidiophore with chains of conidia. **G,** *Alternaria:* Conidiophore with a chain of three conidia, spores of four different species. **H,** *Chaetomium:* **1,** ascus with eight ascospores, **2,** ascospores more highly magnified, **3,** perithecium. (A, C, F, and H-3 after Barnett; E-1 after Wolf; E-2 after Ellis.)

or two germ tubes. The conidiophores are septate, varying from 19.2 to 105.6 μm long. The hyphae are septate, profusely branched, and somewhat raised.

Disease Cycle

The fungus is seedborne and soilborne. Infection occurs directly or through wounds in stalk tissue. Whether this is an active, aggressive pathogen is questionable.

Other Stalk Infections

Numerous other fungi cause maize stalk infections, but are normally of little economic importance.

Nigrospora oryzae (Berk. & Br.) Petch. This fungus causes black, shallow lesions on the stalks as plants approach maturity. It is most common on ears, husks, and shanks (see ear and kernel rots, page 51). *Nigrospora* is usually a secondary invader.

Pyrenochaeta terrestris (Hansen) Gorenz, Walker & Larson. *Pyrenochaeta* commonly infects the basal portions of maize stalks, primarily those below the soil surface. Shallow, dark brown, blotchy lesions become blended with reddish areas as the plants mature (Fig. 55). The pycnidia are subglobose, ostiolate, dark brown to black with scattered setae about the ostiole. The spores are hyaline, oblong to ovoid, 3.5–5.8 × 1.8–2.3 μm, and sessile within the pycnidia (Fig. 53C).

Phaeocytostroma ambiguum (Mont.) Petr. (Syn. *Phaeocytosporella zeae* Stout). This fungus causes straw-colored, oblong blotches close to the base of stalks that resemble those of Diplodia stalk rot. The spores are one-celled, long-ovoid, brown, 9–15.5 × 4–6 μm, and borne on simple conidiophores in elongated, erumpent pycnidia with a long neck. The disease has been reported from the USA and France.

Physoderma maydis Miy., the causal agent of Physoderma brown spot (page 19), also produces water-soaked lesions on maize stalks beneath the leaf sheath. (See Plate 16.) Lesions may coalesce to form brown blotches. As the disease develops, small pockets of brown sporangia form in the stalks.

Other fungi are generally considered secondary invaders although some can cause stalk rots following

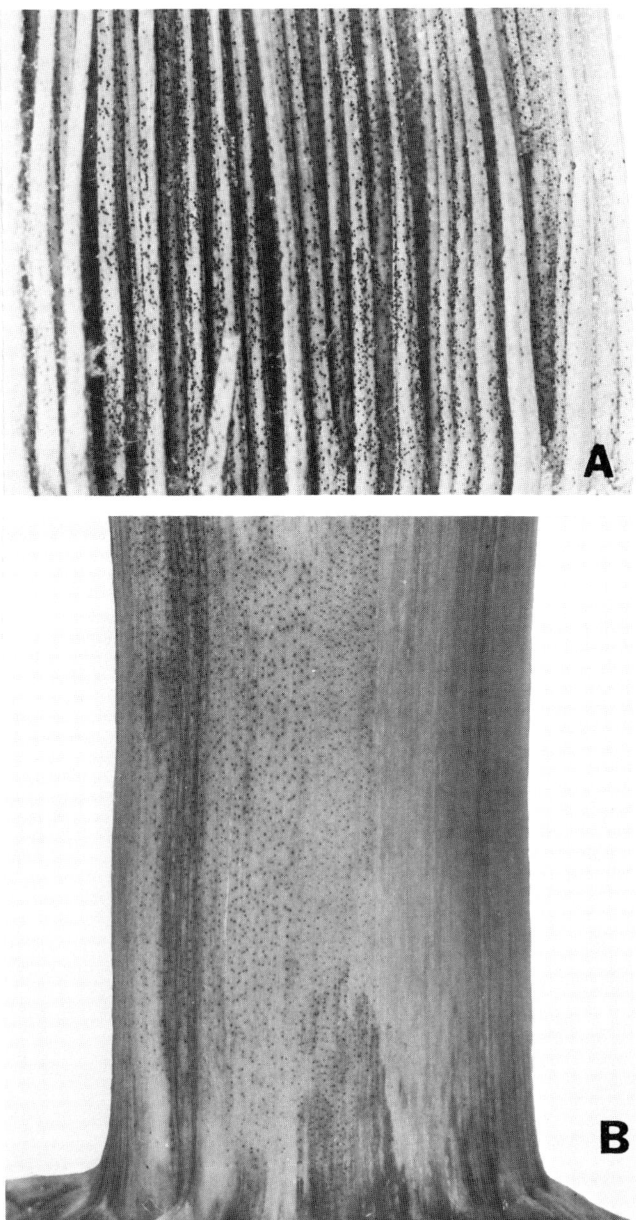

Fig. 54. Charcoal rot is characterized by numerous black sclerotia in **(A)** the stalk's interior visible **(B)** just beneath the stalk surface.

Fig. 55. Pyrenochaeta stalk rot, center and right stalks. Left stalk was infected by *Gibberella zeae* at the upper edge of the brace roots while *Diplodia maydis* started at the root just below the soil line. *Pyrenochaeta terrestris* produces shallow, dark brown, blotchy stalk lesions usually below the soil line. All roots on the stalks above were removed at the stalk surface.

1. Stewart's bacterial wilt. (Courtesy F. M. Latterell)

2. Nebraska bacterial wilt and leaf freckles. (Courtesy M. L. Schuster)

3. Bacterial leaf blight; healthy leaf on top. (Courtesy N. W. Schaad and D. R. Sumner)

4. Chocolate spot. (Courtesy D. C. Arny)

5. Corn stunt—Mesa Central strain. (Courtesy A. J. Ullstrup)

6. Corn stunt—Rio Grande strain. (Courtesy R. Gamez)

7. Maize bushy stunt—early and late symptoms. (Courtesy L. R. Nault)

8. Maize bushy stunt. (Courtesy L. R. Nault)

9. Diplodia seedling blight.

10. Southern corn leaf blight, Race O. (Courtesy R. R. Bergquist)

11. Southern corn leaf blight, Race T. (Courtesy D. C. Arny)

12. *Helminthosporium maydis*, Race T, infection over an ear shank. (Courtesy D. C. Arny)

13. Northern leaf blight showing susceptible and resistant type lesions. (Courtesy R. R. Bergquist)

14. Helminthosporium leaf spot—susceptible and resistant seedlings. (Courtesy A. J. Ullstrup)

15. Physoderma brown spot. (Courtesy W. G. Willis)

16. Physoderma stalk infection. (Courtesy R. R. Nelson)

17. Yellow leaf blight. (Courtesy G. L. Worf)

18. Eyespot—backlighted. (Courtesy G. L. Worf)

19. Close-up of anthracnose lesions on a leaf blade. (Courtesy R. R. Bergquist)

20. Phaeosphaeria leaf spot. (Courtesy A. J. Ullstrup)

21. Cercospora or gray leaf spot. (Courtesy R. E. Stuckey)

22. Zonate leaf spot on iron-deficient leaf. (Courtesy A. J. Ullstrup and B. L. Renfro)

23. Tar spot. (Courtesy Carlos deLeon)

24. Septoria leaf blotch. (Courtesy Carlos deLeon)

25. Banded leaf and sheath spot. Note sclerotia and mycelium. (Courtesy B. L. Renfro)

26. Diplodia leaf streak. (Courtesy F. M. Latterell)

27. Hyalothyridium leaf spot. (Courtesy F. M. Latterell)

28. Brown stripe downy mildew. (Courtesy A. J. Ullstrup and B. L. Renfro)

29. Sugarcane downy mildew. (Courtesy A. J. Ullstrup and B. L. Renfro)

30. Philippine downy mildew. (Courtesy Carlos deLeon)

31. Head smut on ear. (Courtesy A. J. Ullstrup)

32. Common maize rust. (Courtesy Carlos deLeon)

33. Southern rust. (Courtesy R. R. Bergquist)

34. Tropical rust. (Courtesy Carlos deLeon)

35. Top dieback symptoms of anthracnose. (Courtesy R. E. Stuckey)

36. Gibberella stalk rot. (Courtesy T. M. Sjulin)

37. Fusarium stalk rot. (Courtesy R. R. Bergquist)

38. Pythium stalk rot. (Courtesy B. L. Renfro)

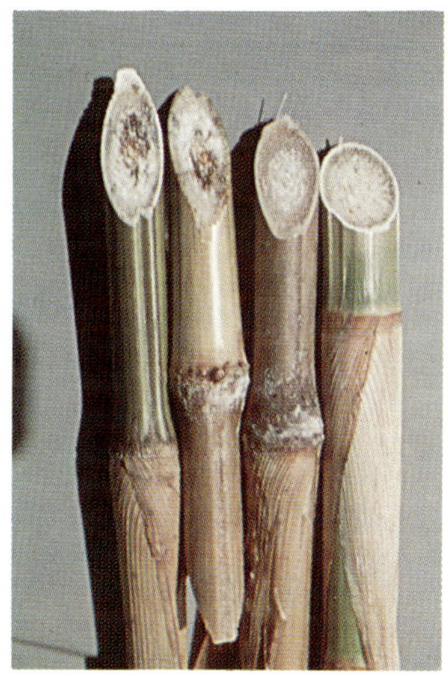

39. Cephalosporium stalk rot (late wilt). (Courtesy A. J. Ullstrup and B. L. Renfro)

40. Three common ear rots: **A,** Diplodia, **B,** Fusarium, **C,** Gibberella.

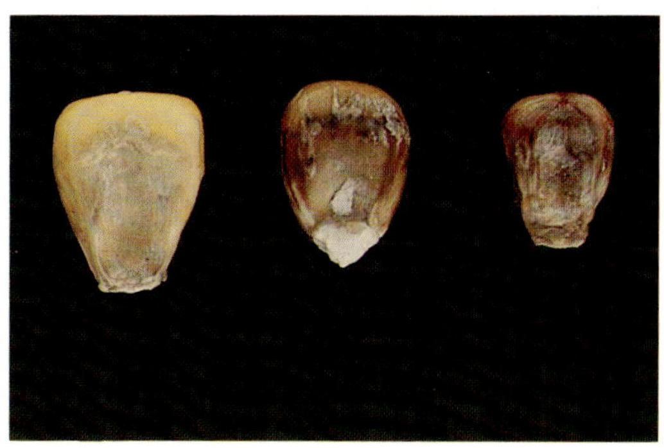

41. Three kernels infected with *Gibberella zeae.* (Courtesy John Tuite)

Guide to Nutrient Deficiency Symptoms

HEALTHY leaves shine with a rich dark green color when adequately fed

PHOSPHATE shortage marks leaves with reddish-purple, particularly on young plants.

POTASH deficiency appears as a firing or drying along the tips and edges of lowest leaves.

NITROGEN hunger sign is yellowing that starts at tip and moves along middle of leaf.

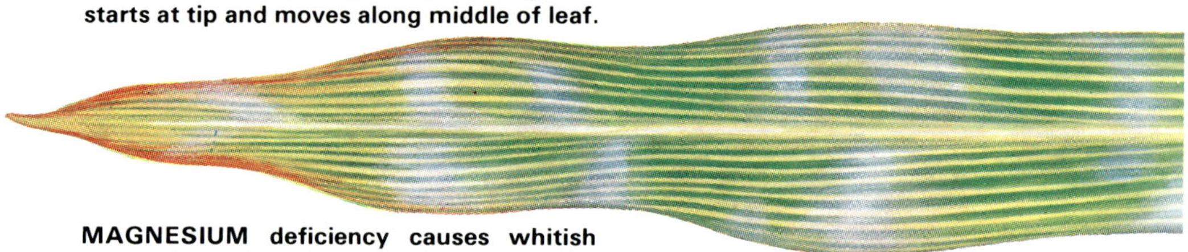

MAGNESIUM deficiency causes whitish strips along the veins and often a purplish color on the underside of the lower leaves.

DROUGHT causes the corn to have a grayish-green color and the leaves roll up nearly to the size of a pencil.

Drawings: Maynard Reece

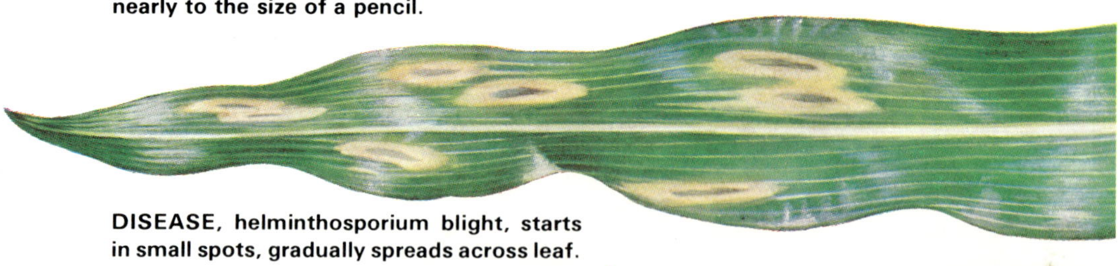

DISEASE, helminthosporium blight, starts in small spots, gradually spreads across leaf.

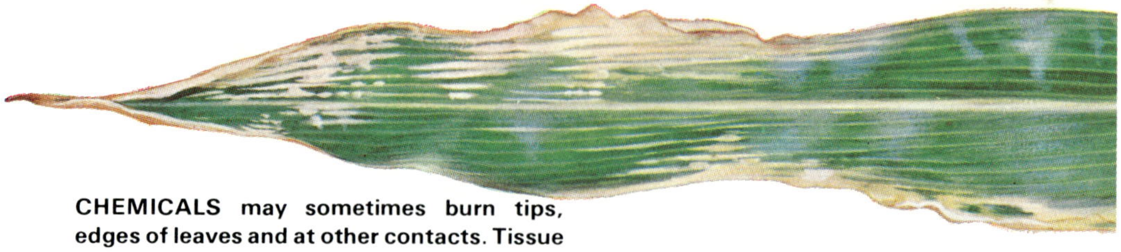

CHEMICALS may sometimes burn tips, edges of leaves and at other contacts. Tissue dies, leaf becomes whitecap.

DEEP, SPREADING ROOTS of healthy, high-yielding plant will crowd a half-bushel basket.

PHOSPHATE shortage during early weeks causes a shallow root system with little spread.

ROOTWORMS prune heavily as they eat small roots, tunnel in larger ones.

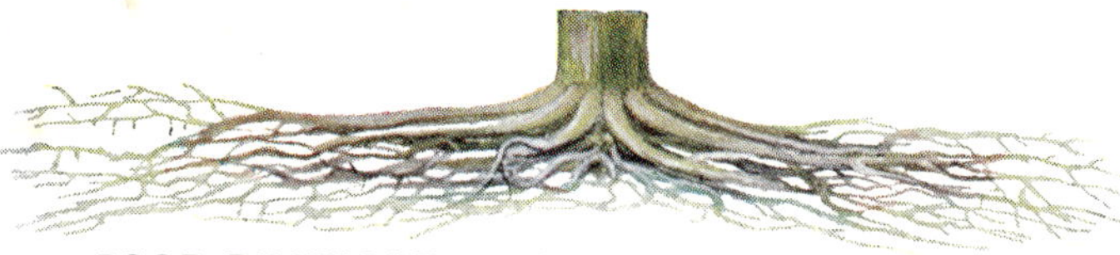

POOR DRAINAGE and hardpan are causes of a flat, shallow root system. Corn with poor roots can't stand drought and is easily blown over by high winds.

ACID SOIL is indicated when the lower part of the root is discolored and decayed, particularly when brace roots shoot from third or fourth node.

↑**PRUNED ROOTS** are work of a cultivator. Shovels were too deep and too close.

CHEMICAL damage makes roots writhe and twist. Joined brace roots are another symptom. ↓

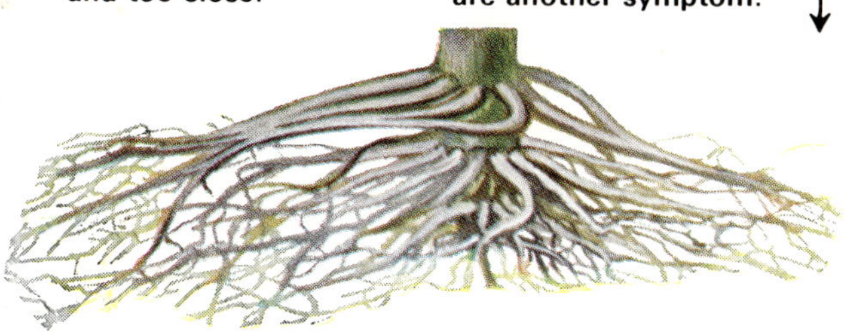

NORMAL EAR on well fertilized high-producing corn weighs about 2/3 lb. It has well filled tips.

BIG EARS weighing up to 1 lb. indicate that plant population was too small for most profitable yields.

SMALL EARS usually are a sign of low fertility. For better yields, boost fertilizer application.

POTASH shortage shows up in ears with poorly filled tips and loose chaffy kernels.

PHOSPHATE shortages interfere with pollination and kernel fill. Ears are small, often are twisted and with undeveloped kernels.

NITROGEN is essential throughout the growing season. If plant runs out of nitrogen at critical time, ears are small and protein content is low. Kernels at tip do not fill.

GREEN SILKS at maturity may be caused by too much nitrogen in relation to other elements.

DRY WEATHER slows silking behind tasseling; kernels aren't pollinated.

42. Penicillium ear rot. (Courtesy A. J. Ullstrup and B. L. Renfro)

43. Aspergillus ear rot caused by a member of the *Aspergillus glaucus* group. (Courtesy John Tuite)

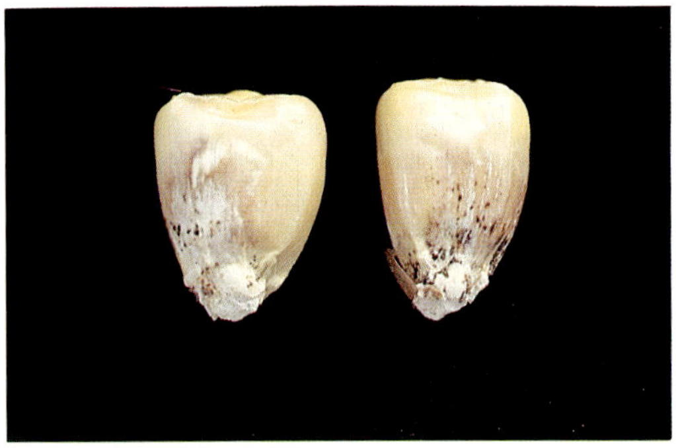

44. Two kernels infected with the anthracnose fungus (*Colletotrichum graminicola*). (Courtesy H. L. Warren)

45. Ear rot caused by *Helminthosporium maydis*, Race T. (Courtesy D. C. Arny)

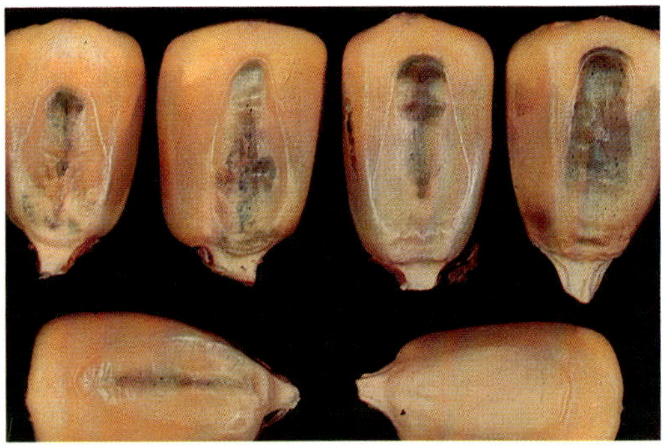

46. "Blue-eye" kernels infected with *Penicillium*. Kernel at lower right is healthy.

47. Aspergillus storage mold. (Courtesy John Tuite)

48. Maize dwarf mosaic (sugarcane mosaic). (Courtesy B. L. Renfro)

49. Maize chlorotic dwarf; healthy plant on right. (Courtesy V. D. Damsteegt)

50. Maize chlorotic mottle. (Courtesy J. Castillo L.)

51. Rayado fino. Healthy leaf at left. (Courtesy R. Gamez)

52. Initial symptoms of maize streak on very young leaves. (Courtesy V. D. Damsteegt)

53. Typical maize streak. (Courtesy V. D. Damsteegt)

54. Close-up of corn lethal necrosis. Note severe chlorosis of affected plant. (Courtesy J. K. Uyemoto)

55. Dagger (*Xiphinema*) nematode damage in a field of corn. (Courtesy B. J. Jacobsen)

56. Severe stubby-root (*Trichodorus*) nematode damage to corn roots. Note early tassel formation. (Courtesy R. S. Hussey)

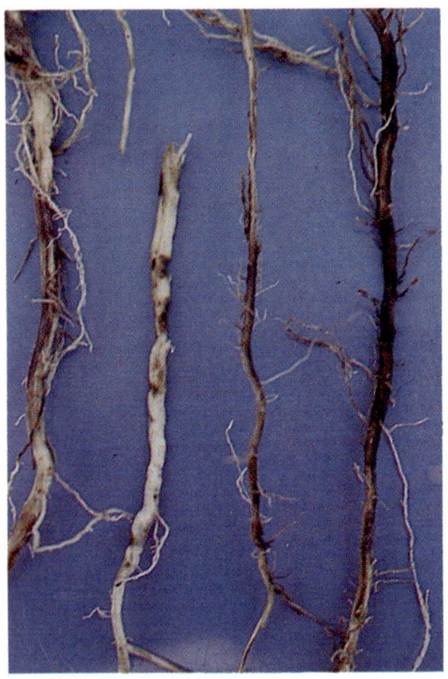

57. Dark, discolored areas in roots due to feeding of root-lesion (*Pratylenchus*) nematodes.

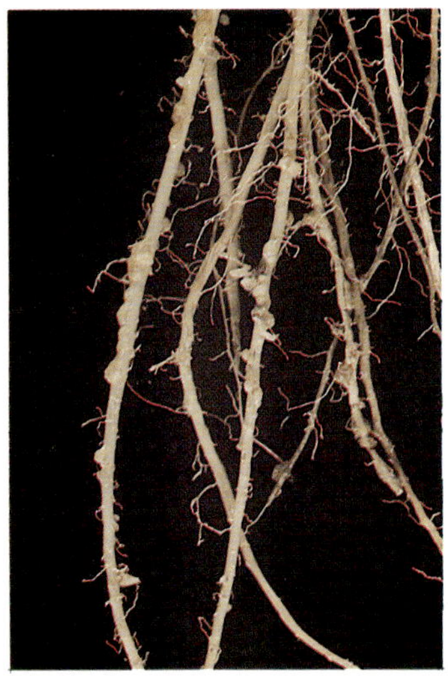

58. Root-knot (*Meloidogyne*) injury. (Courtesy E. Brown)

59. *Striga* (witchweed) parasitizing a corn plant. (USDA photo)

60. Anhydrous ammonia burn. (Courtesy D. C. Arny)

61. Cyanazine (Bladex) injury. (Courtesy E. L. Knake)

62. Butylate (Sutan) injury. (Courtesy B. J. Jacobsen)

63. Linuron (Lorox) injury. (Courtesy J. P. Downs)

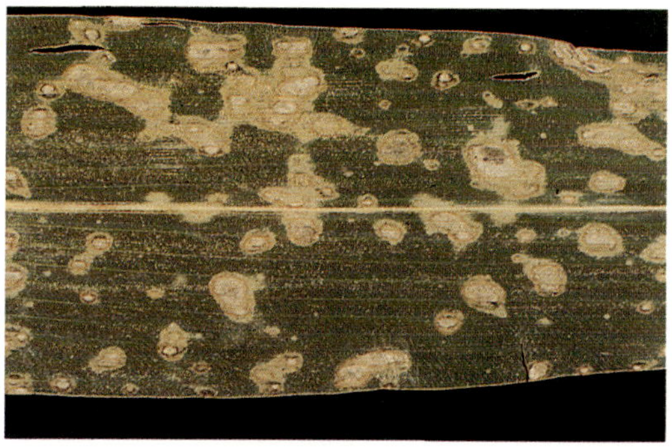

64. Bipyridilium (Paraquat) injury. (Courtesy F. M. Latterell)

65. Trifluralin (Treflan) injury. (Courtesy R. R. Bergquist)

66. Zinc deficiency. (Courtesy J. R. Jared)

67. Drought stress. (Courtesy W. O. Scott)

68. Flood damage. (Courtesy W. O. Scott)

69. Frost injury. (Courtesy D. C. Arny)

70. High-temperature sunscald.

71. Cold banding of young leaves. (Courtesy D. C. Arny)

72. Driving winds and sand particles caused this severe injury. (Courtesy W. O. Scott)

73. Red coloration from broken midrib.

74. Ozone damage. (Courtesy A. Clyde Hill)

75. Fluoride injury; healthy leaf at left. (Courtesy A. H. Epstein)

76. Chlorine injury.

artificial inoculations. These include species of *Alternaria* (Fig. 53G), *Aspergillus* (Fig. 57B), *Botrytis* (Fig. 53E), *Chaetomium* (Fig. 53H), *Cladosporium* (Fig. 57E), *Curvularia* (Fig. 25), *Fusarium* (Fig. 51), *Helminthosporium* (Fig. 17), *Ligniera*, *Mucor* (Fig. 53D), *Penicillium* (Fig. 57C), *Physalospora* (Fig. 57D), *Rhizoctonia* (Fig 57A), *Rhizopus* (Fig. 57I), *Spicaria* (Fig. 53F), and *Trichoderma* (Fig. 57G).

General Epidemiological Features of Fungal Stalk Rots

1. Fungal stalk rots are generally more severe when N is in excess in relation to K.
2. Leaf blights, hail, or insect damage reduce leaf area and increase the potential for stalk rot.
3. Evidence shows that maize plants with two ears are the first to be infected.
4. Insects (e.g., European corn borer, southern and northern corn rootworms) may aid in the development of stalk rots by carrying inoculum into tissues, causing wounds through which pathogens enter stalks or roots, and reducing the photosynthetic area.
5. Genetic resistance involves multiple genes, some with major effects.
6. Early maturing hybrids are generally more susceptible than full-season varieties.
7. High plant populations and narrow rows increase stalk rot potential.
8. Severe stalk rots may involve several organisms.

Control

1. Select resistant hybrids and varieties adapted to the area.
2. Plant sound, disease-free seed treated with a fungicide.
3. Attain balanced soil fertility, especially between N and K, by applying fertilizer based on soil test results.
4. Adjust the plant population to the particular hybrid or variety, the fertility level, soil type, and available soil moisture in the field.
5. Crop rotation and a clean plowdown of crop residues may reduce stalk rot in some cases.
6. Control root-attacking and stalk-attacking insects using approved resistant varieties, cultural practices, and chemicals where needed and feasible.
7. Harvest when the crop is mature (30% grain moisture) to prevent losses from lodging.
8. Where practical, irrigate during droughts until 50–55 days after flowering.

Selected References

CHRISTENSEN, J. J., and R. D. WILCOXSON. 1966. Stalk Rot of Corn. Monograph No. 3. Am. Phytopathol. Soc.: St. Paul, MN.

DODD, J. L. 1977. A photosynthetic stress-translocation balance concept of corn stalk rot, pp. 122-130. Proc. 32nd Annu. Corn and Sorghum Res. Conf. Am. Seed Trade Assoc.: Washington, DC.

ELLIOTT, C. 1943. A Pythium stalk rot of corn. J. Agric. Res. 66:21-39.

HARRIS, M. R. 1936. The relationship of *Cephalosporium acremonium* to the black-bundle disease of corn. Phytopathology 26:965-980.

KOEHLER, B. 1960. Cornstalk rots in Illinois. Ill. Agric. Exp. Stn. Bull. 658.

KUCHAREK, T. A., and T. KOMMEDAHL. 1966. Kernel infection and corn stalk rot caused by *Fusarium moniliforme*. Phytopathology 56:983-984.

LATTERELL, F. M., A. E. ROSSI, and R. MORENO. 1976. *Diplodia macrospora*: A potentially serious pathogen of corn in the US? (Abstr.). Proc. Am. Phytopathol. Soc. 3:228.

NELSON, P. E., T. A. TOUSSOUN, and R. J. COOK. 1980. Fusarium: Biology and Taxonomy. Penn State Univ. Press: University Park.

SAMRA, A. S., K. A. SABET, and M. K. HINGORANI. 1963. Late wilt disease of maize caused by *Cephalosporium maydis*. Phytopathology 53:402-406.

Ear and Kernel Rots

Maize is susceptible to a number of ear and kernel rots, some of which are widely distributed. These rots cause considerable damage in humid areas, especially when rainfall is above normal from silking to harvest. The prevalence of rots can be increased by insect and bird damage to ear and stalk and by lodging where ears touch the soil. Ears well covered by husks and maturing in a downward position usually have less rot than ears with open husks or maturing upright. Ear and kernel rots can reduce yield, plus the quality and feed value of grain.

Diplodia Ear Rots

Symptoms

The husks of early infected ears appear bleached or straw-colored. If infection occurs within two weeks after silking, the entire ear turns grayish brown, shrunken, is very lightweight, and completely rotted. Lightweight ears usually stand upright with the inner husks adhering tightly to one another or the ear because of mycelial growth between them. Black pycnidia may be scattered on the husks, floral bracts, cob tissues, and the sides of kernels. Ears infected later in the growing season show no external symptoms, but when ears are broken and kernels removed, a white mold is commonly found growing between the kernels whose tips are discolored. Infection usually begins at the ear base, moving up from the shank (Plate 40A). Some isolates of *D. maydis* induce vivipary (premature germination).

Causal Organisms

Diplodia maydis and *D. macrospora* (see stalk rots, page 43).

Disease Cycle

See Diplodia stalk rot, page 44.

Epidemiology

Dry weather early in the season followed by wet conditions just before and after silking favors ear infection. Ears are most susceptible from silking until about three weeks later. Hybrids with thin pericarps are often very susceptible. Bird and insect feeding on the ear tips also increases the potential for ear rot.

Control

1. Resistant hybrids and varieties.
2. Early harvest.
3. Proper storage (below 18% moisture initially for ears, 15% for shelled grain) will prevent further growth of the pathogen, but is not a control measure in the field. See page 57, control of storage rots.

Fusarium Kernel or Ear Rot

Symptoms

A salmon-pink to reddish-brown discoloration first appears on the caps of individual kernels or groups of kernels scattered over the ear. As the disease progresses, a powdery or cottony-pink mold growth develops on the infected kernels. The fungus growth commonly becomes established around channels made by corn earworms or corn borers. This is probably the most widespread disease of maize ears (Plate 40B). Kernels infected late in the season develop whitish streaks on the pericarp.

Causal Organisms

Fusarium moniliforme and *F. moniliforme* var. *subglutinans* (see stalk rots, page 43).

Disease Cycle

See Fusarium stalk rot, page 45).

Epidemiology

Disease development and spread are favored by dry, warm weather. Infection may follow injury from growth cracks or other pericarp injuries, bird feeding, corn earworms, corn borers, or other insects. Certain high-lysine maize hybrids may be more susceptible to *F. moniliforme* ear rot.

Control

See Diplodia ear rots (above).

Gibberella or Red Ear Rot

Symptoms

A reddish mold, often beginning at the ear tip, is the characteristic sign of Gibberella ear rot. Early infected ears may rot completely with the husks adhering tightly to the ear and a pinkish to reddish mold growing between the husks and ear. Superficial, blue-black perithecia are sometimes found on the husks and ear shanks (Plates 40C and 41). Gibberella-infected ears are toxic to swine and other animals (see Table IV).

Causal Organism and Disease Cycle

Gibberella zeae (see Gibberella stalk rot, page 45).

Epidemiology

Cool, wet weather within three weeks of silking favors development of this disease.

Control

See Diplodia ear rots (above).

Nigrospora Ear or Cob Rot

Symptoms

Symptoms are not conspicuous until the ears are harvested. The distinguishing feature is a shredding of the cob, usually at the butt end. Shredding occurs when ears are picked mechanically or later when the ears are shelled. Kernels may show a gray mycelium and are "peppered" with masses of small, round, black spores (Fig. 56A). Affected ears are lightweight and chaffy, with kernels slightly bleached or streaked (Fig. 56B), poorly finished, and easily pressed into the cob. In yellow and white hybrids, the chaff is brown and pale yellow or gray, respectively. The disease is widely distributed throughout the world.

Causal Organism

Nigrospora oryzae (Berk. & Br.) Petch. (Syn. *Basisporium gallarum* Möll.). Perfect state is *Khuskia oryzae*.

Spores are black, spherical to ovoid, 10–16 μm (mostly 12–14 μm) in diameter, and borne singly on short, branched conidiophores, 3–7 μm thick (Fig. 57A). Conidiophores are erect with short branched apices.

Disease Cycle

The fungus overseasons on plant debris in the field, particularly in undeveloped secondary ears. *N. oryzae* is a weak parasite and attacks ears only after plants are weakened or killed prematurely by frost, drought, root injury, leaf blights, or other stalk rot fungi.

Fig. 56. Nigrospora ear and cob rot. Black dots at kernel tips are spore masses (top). Whitish streaks are due to air channels in the pericarp (bottom).

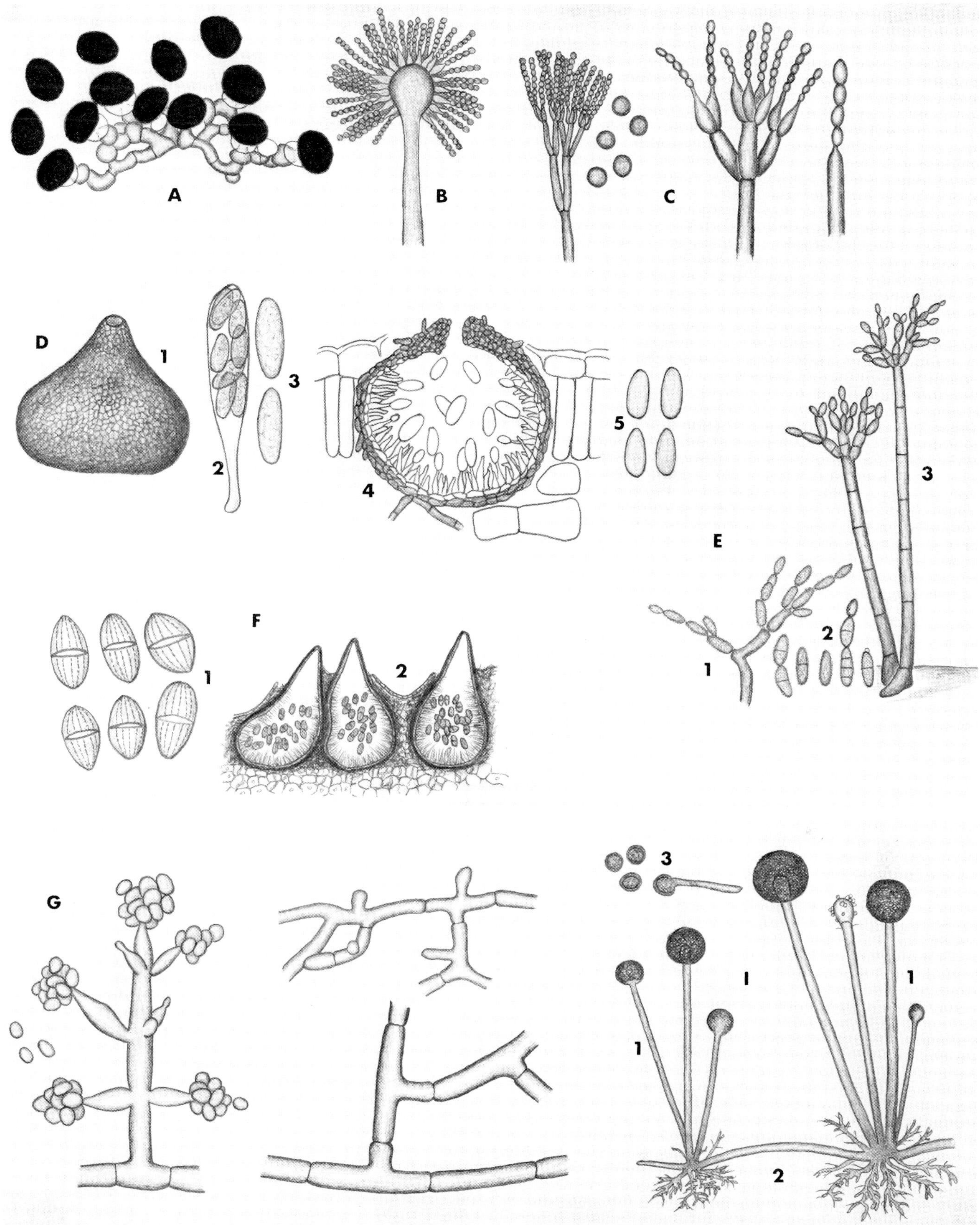

Fig. 57. Fungi causing ear and kernel rots. **A,** *Nigrospora oryzae:* conidiophores and conidia. **B,** *Aspergillus:* conidiophore with chains of conidia. **C,** several species of *Penicillium* showing three conidiophores with chains of conidia and spores more highly magnified. **D,** *Physalospora* or *Botryosphaeria:* **1,** perithecium, **2,** ascus with eight ascospores, **3,** two ascospores more highly magnified, **4,** section through a pycnidium containing pycnidiospores, **5,** four pycnidiospores more highly magnified. **E,** *Cladosporium herbarum:* **1,** part of a conidiophore with conidia, **2,** conidia, **3,** conidiophores with conidia. **F,** *Botryodiplodia:* **1,** mature pycnidiospores; **2,** section through a stroma showing three pycnidia containing pycnidiospores. **G,** *Trichoderma viride:* conidiophore with clusters of conidia. **H,** mycelium of *Rhizoctonia.* **I,** *Rhizopus* showing **1,** sporangiophores and sporangia of various ages, **2,** stolon and rhizoids, **3,** several spores more highly magnified, one germinating. (A, E-1,2 after Ellis; D-4,5 and E-3 after Barnett; F-2 after Saccas.)

Epidemiology

Maize grown in infertile soil appears to be more susceptible than that grown in fertile soil, possibly because poor nutrition causes premature dying. Stalk or root rots, leaf blights, hail, insect injury, and other damage increase disease severity.

Control

1. Full-season, adapted hybrids and varieties resistant to stalk rots and leaf blights.
2. Balanced soil fertility.
3. Control of root-feeding and stalk-feeding insects.
4. See Diplodia ear rots (above).

Gray Ear Rot

Symptoms

The grayish-white mold on and between the kernels near the ear base resembles that of Diplodia ear rots (above). Early infected ears have tightly adherent and bleached husks. At harvest, such ears stand upright and are slate gray instead of grayish brown as in Diplodia ear rots. Small black sclerotia are often scattered throughout the cob (Fig. 58). Severely infected kernels have slate gray to black streaks or specks under the pericarp.

Causal Organism

Botryosphaeria zeae (Stout) von Arx & Müller (Syn. *Physalospora zeae* Stout), asexual state *Macrophoma zeae* Tehon & Daniels

Fig. 58. Gray ear rot caused by *Physalospora zeae.* In early stages, infection may resemble Diplodia ear rot as in ear at left. In final stages (right), ears are shriveled and black.

Perithecia and pycnidia occur in large lesions on corn leaves and occasionally on tassel necks and under the uppermost leaf sheath. Perithecia are black and contain clavate to cylindrical asci measuring 17–22 × 85–175 μm. Ascospores are hyaline to greenish, oval to ellipsoid, one-celled, and 8–11 × 21–33 μm.

The pycnidia of *M. zeae,* which may develop in the same lesion with the perithecia, are submerged, black, globose, and protruding with short necks. They produce unicellular, ellipsoid conidia, 6.5–8.5 × 17–31 μm (Fig. 57D).

Disease Cycle

The fungus overseasons on infected leaves, and the spores mature the following growing season. They are spread by wind and rain to infect susceptible ears and leaves. Perithecia and pycnidia do not develop in the ears. Only sclerotia are found in rotted kernels and ears. They serve as a means of survival and propagation for the causal fungus.

Epidemiology

Extended periods of warm-to-hot, wet weather for several weeks after silking favor disease development. The disease is unimportant.

Control

Adapted, resistant hybrids and varieties. Resistance to gray and Diplodia ear rots may be correlated.

Penicillium Rot

Symptoms

Penicillium rot occurs primarily on ears injured mechanically or by insects. The characteristic sign is the powdery, green or blue-green mold on and between the kernels, usually at the ear tip (Plate 42). "Blue-eye" rot occurs on stored grain with high moisture content. (See storage rots, page 56.) Kernels with fungal growth normally become bleached and streaked.

Causal Organisms

Penicillium oxalicum Currie & Thom is the most common, although other species, including *P. chrysogenum, P. glaucum,* and *P. funiculosum,* may commonly occur.

Penicillium species are soil-inhabiting fungi with long conidiophores that branch in a broomlike fashion. The conidia are globose to ovoid, and under the microscope resemble glass beads (Fig. 57C). *Penicillium* spp. can invade seed embryos and cause seedling blights on germination.

Black Kernel Rot

Symptoms

During early stages of infection, kernels show a brown discoloration. As the disease progresses, the kernels become black with small eruptions that may contain pycnidia. A dry rot develops later.

Causal Organism

Botryodiplodia theobromae Pat.

Fungus growth on PDA is white and aerial with a septate, branched mycelium and oval, brown, one-septate conidia measuring 10.5–17.25 × 21.5–31.5 μm. Black, subglobose to globose pycnidia may occur singly or in clumps on the medium (Fig. 57F).

Cladosporium Rot

Symptoms

Cladosporium kernel or ear rot is identified by dark, greenish-black, blotched or streaked kernels scattered over the ear (Fig. 59). Discoloration appears first where the kernels are attached to the cob. Later, blotches appear upward but seldom reach the crown. Further rotting occurs during storage. When completely colonized, the ears are dark and lightweight. The disease has not been reported to be of economic importance.

Causal Organism

Cladosporium herbarum Lk. ex Fr. and *C. cladosporioides* (Fres.) De Vries (*Hormodendrum cladosporioides* (Fres.) Sacc.)

Conidiophores are erect, up to 250 μm long, smooth, sparsely branched, and pale olive-brown to brown. The conidia often occur in chains of two or three, are pale olive-brown, one to three septate, subcylindrical, and 5–23 × 3–8 μm (mostly 8–15 μm) (Fig. 57E). The fungi can invade the crowns of kernels damaged by growth cracks.

Physalospora (Botryosphaeria) Ear Rot

Symptoms

Physalospora ear rot develops as a dark brown to black, felty mold growth on all parts of the ear. Mildly infected ears may have a few blackened kernels near the base (Fig. 60).

Causal Organism

Botryosphaeria festucae Lib. von Arx & Müller (Syn. *Physalospora zeicola* Ell. & Ev.), asexual state *Diplodia frumenti* Ell. & Ev.

The black, submerged perithecia are gregarious with conical necks that emerge through the cuticle. Asci are clavate-cylindrical, double-walled, and measure 10–13 × 95–140 μm. The ascospores, usually eight and arranged biseriately, are hyaline, unicellular, oval, and 8–9 × 20–30 μm (Fig. 57D). The pycnidia are submerged, black, and contain oval, two-celled, dark brown, striate conidia. Fruiting structures of both stages are produced on old stalks and husks. Warm, humid weather seems favorable for infection.

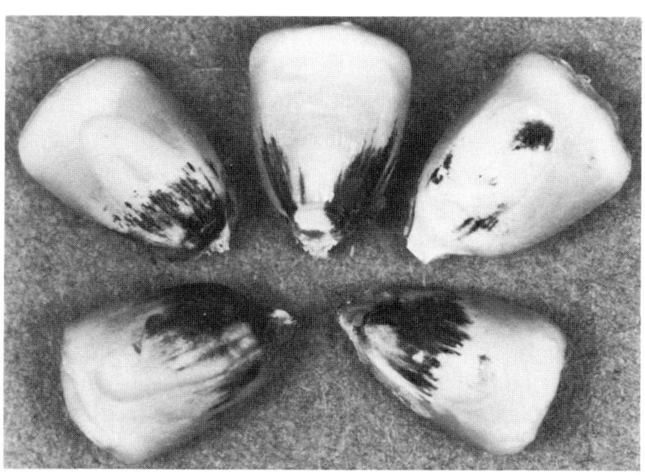

Fig. 59. Cladosporium or Hormodendrum infection of maize kernels.

Rhizoctonia Ear Rot

Symptoms

Rhizoctonia ear rot is recognized in its early stages by a salmon-pink mold growth on the ear. Later, infected ears become dull gray with numerous, white to salmon-colored to dark brown or black sclerotia developing on the outer husks (Fig. 61).

Causal Organism

Rhizoctonia zeae Voorhees

R. zeae overseasons as dormant mycelium and

Fig. 60. Physalospora (Botryosphaeria) ear rot caused by the fungus *Physalospora zeicola*. A dark brown to black felty mold covers the entire ear in severe infections.

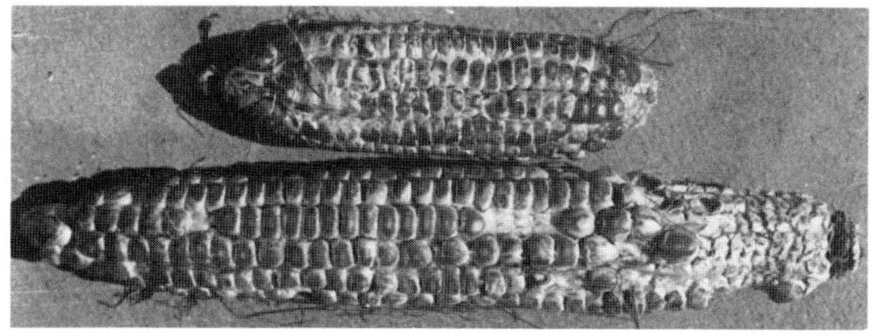

Fig. 61. Rhizoctonia ear rot. (Courtesy B. L. Renfro)

Fig. 62. Rhizopus ear rot.

Fig. 63. Trichoderma ear rot.

sclerotia in kernels, soil, or on and in plant debris. The sclerotia propagate and disseminate the fungus and are mostly superficial, 0.1–0.5 mm in diameter. Warm, humid weather appears favorable for infection. *Rhizoctonia* hyphae are typically constricted at right angles (Fig. 57H). No spores are produced.

Other Ear Rots

Several species of *Aspergillus* (Fig. 57B and Plate 43) and *Rhizopus* (Fig. 62) on maize ears have been described. *Aspergillus* spp. cause black or greenish-yellow or tan growth on and between the kernels (Plate 43); *Rhizopus* spp. produce a coarse, white mold growth with numerous minute sporangia appearing as black specks (Fig. 62). Ears can also be infected with *Colletotrichum graminicola* (Plate 44). *Trichoderma viride* Pers. ex Fr. (Figs. 57G and 63) produces green, fuzzy mycelia on and between the kernels, often developing after severe leaf damage by other fungi or *Helminthosporium maydis* (Race T) ear infection (Plate 45).

Helminthosporium maydis, H. rostratum, H. carbonum and other species may infect maize ears (see Helminthosporium leaf spots, page 17 and Fig. 17).

Selected References

BEZUIDENHOUT, H., and W. F. O. MARASAS. 1978. *Botryosphaeria zeae*: The cause of grey ear rot of maize (*Zea mays*) in South Africa. Phytophylactia 10:21-24.

CALVERT, O. H., M. S. ZUBER, and M. G. NEUFFER. 1969. Vivipary in *Zea mays* induced by *Diplodia maydis*. Phytopathology 59:239-240.

DEL ROSARIO, M. S. 1954. The fungus causing black kernel rot of corn. Philipp. Agric. 37:623-631.

DHANRAJ, K. S. 1966. Dry rot of maize caused by *Diplodia macrospora* Earle. Indian Phytopathol. 19:120.

KOEHLER, B. 1959. Corn ear rots in Illinois. Ill. Agric. Exp. Stn. Bull. 639.

MOHAMED, H. A., and S. M. FATHI. 1965. Nigrospora cob rot of corn in the United Arab Republic. Plant Dis. Rep. 49:244-246.

WARREN, H. L. 1978. Comparison of normal and high-lysine maize inbreds for resistance to kernel rot caused by *Fusarium moniliforme*. Phytopathology 68:1331-1335.

Storage Rots

Storage rots may develop on ear or shelled maize in storage, causing a reduction in feed and market value. Certain storage-rotting fungi produce toxins or hormones toxic to animals, but this is apparently uncommon. Materially discolored kernels (including those with discolored germs) reduce the crop's market value.

Symptoms

The first stages of kernel invasion by fungi are not detectable by the unaided eye. External symptoms appear as mold growth on or between the kernels after damage is fairly well advanced. Invasion of whole kernels in bins results in discoloration, heating, caking, and mustiness (Fig. 64). One storage rot, called "blue-eye," is characterized by a bluish-green germ (Plate 46).

Causal Organisms

Storage rots are caused by many fungi, principally species of *Aspergillus* (Fig. 57B and Plate 47) and

Fig. 64. Discoloration of kernels from infection by storage fungi. (Courtesy C. M. Christensen)

Penicillium (Fig. 57C and Plates 42 and 46). They occur alone or in combination with other fungi. Some species of *Fusarium* often overrun kernels with ruptured pericarps.

Epidemiology

Storage fungi thrive in shelled maize with moisture contents of 15–20% and at 21–32° C. At moisture contents below 15% and temperatures below 10° C, the risk of damage by storage fungi is very low.

Control

1. Dry shelled maize to 13–15% moisture and, where possible, aerate to maintain a uniform temperature of 4–10° C throughout the bulk.
2. Chemical storage additives are being used to a limited extent (see mycotoxins, below).
3. Central American maize hybrids and varieties show evidence of resistance.

Selected References

CHRISTENSEN, C. M., and H. H. KAUFMANN. 1968. Maintenance of quality in stored grain and seeds. Univ. Minn. Agric. Ext. Serv. Folder 226.

CHRISTENSEN, C. M., and H. H. KAUFMANN. 1969. Grain Storage: Role of Fungi in Quality Loss. Univ. Minn. Press: Minneapolis, MN.

CHRISTENSEN, C. M., and H. H. KAUFMANN. 1974. Microflora, pp. 158-192. In: CHRISTENSEN, C. M. (ed.). Storage of Cereal Grains and Their Products. Am. Assoc. Cereal Chem.: St. Paul, MN.

QASEM, S. A., and C. M. CHRISTENSEN. 1960. Influence of various factors on the deterioration of stored corn by fungi. Phytopathology 50:703-709.

TUITE, J., and G. H. FOSTER. 1979. Control of storage diseases of grain. Annu. Rev. Phytopathol. 17:343-366.

Mycotoxins and Mycotoxicoses

Mycotoxins are fungal metabolites that are toxic when consumed by animals or man. Mycotoxins can accumulate in maturing corn in the field and in grain during transportation and storage under conditions of moisture, humidity, and temperatures favorable for growth of the toxin-producing fungus or fungi. Diseases in animals and man resulting from the consumption of mycotoxins are termed mycotoxicoses. The effects range from loss of appetite, feed refusal, and decreased feed efficiency to mortality in domestic animals (Table IV). For example, in 1934 in the midwestern USA, more than 5,000 horses died due to "moldy corn disease." In 1972, Gibberella ear rot caused extensive feed-refusal problems in swine in the Corn Belt of the USA.

An outbreak of Turkey X disease in Great Britain in 1960 was traced to contaminated peanut meal from Brazil. Aflatoxin was indicated as the cause of the deaths of over 100,000 young turkeys and some 20,000 ducklings, pheasant and partridge poults. This outbreak stimulated a huge research effort in the USA and increased efforts in countries such as Japan and the USSR.

Aflatoxin B_1, present in corn, is among the most potent of known carcinogens. Ten to 20 parts per billion (ppb) of aflatoxin consumed regularly by sensitive young animals can result in fatal liver cancer, whereas in older animals only minimal effects may occur. No animals are known to be immune. Effects of aflatoxin as a toxin or carcinogen have been demonstrated experimentally in a wide variety of domestic and experimental animals and in human patients who inadvertently consumed contaminated peanut meal. Field outbreaks of diseases have been observed in turkeys, ducks, chickens, swine, cattle, dogs, and trout.

Aflatoxin has been implicated in primary liver cancer in humans. An outbreak of hepatitis in India has been linked to moldy corn containing aflatoxin. Aflatoxin has been found in the tissues of children suffering from Reye's syndrome in the Orient. In 1977, 60% or more of the corn grown in the southeastern USA contained at least 20 ppb ($\mu g/kg$) of aflatoxin B_1—the legal maximum level permitted by the US Food and Drug Administration (FDA) for corn in interstate commerce. The aflatoxin metabolite M_1 tolerance for milk is 0.5 ppb. Although the percentage of contaminated grain varies greatly from year to year, the 1977 data indicate the potential seriousness of the problem.

Three genera of fungi—*Aspergillus, Penicillium,* and *Fusarium (Gibberella)*—are most frequently involved in cases of mycotoxin contamination in corn (Table IV). *Aspergillus flavus* produces aflatoxins in corn starting at a moisture content of about 18%, i.e., in equilibrium with 85% relative humidity (0.85 available water) and at temperatures between 12 and 42° C (optimum 25–32° C). Other toxin-producing fungi grow in the 17–40% moisture range and at a wide range of temperatures from below freezing for species of *Penicillium* and *A. fumigatus* up to 45° C. Grain quality and storability are adversely affected by high moisture content, physical damage to the kernels, and the degree that storage fungi have invaded the seed. Fungi may grow well under a given set of conditions but *not* necessarily produce mycotoxins.

Aflatoxins and Aflatoxicoses

Aspergillus flavus and *A. parasiticus* are ubiquitous in soil and on decaying plant material, cause heating and decay of stored grain, and, under certain conditions, commonly invade corn in the field in the southern USA and the southern Corn Belt of the USA. These fungi are serious in subtropical and tropical regions where maize is an important part of the human diet. Under conditions favorable for aflatoxin production, peanuts, cottonseed, and copra are of especially high aflatoxin risk, but under certain conditions aflatoxin is produced in corn as the kernels are developing on the ears. Aflatoxin production by *A. flavus* is more likely following damage to ears by earworms (Plate 43), weevils and other insects, birds (see Fig. 101), hail, early frost, drought stress, windstorms, or other unfavorable weather. Aflatoxin persists under extreme environmental conditions. Roasting, ammoniation at ambient temperatures, and some microbial treatments may sharply reduce but not necessarily eliminate the aflatoxin content. Presently, these treatments have limited application.

Aflatoxicosis in domestic animals results in lesions in the liver and other tissues, reduced body weight gains and poor growth, and mortality. Young animals are more sensitive than older animals. Acute effects of aflatoxin include induced cellular membrane instability and liver and kidney damage. Prolonged feeding of sublethal doses may lead to teratogenic and hepatocarcinogenic effects.

In poultry, besides fatty liver and kidney disorders, leg and bone problems as well as coccidiosis outbreaks can also result. Aflatoxins may cause vaccines to fail, may lower the bird's susceptibility to disease, and may result in less resistance to general stresses. The increased cellular membrane instability is manifested in increased downgrades and condemnations due to hemorrhages and bruises. Less carcass pigmentation is exhibited and egg yolks are paler. Drops in hatchability and reduced production may be noted as well as smaller eggs with shell problems. Growth is restricted and mortality increases, especially in the growing period.

Ochratoxins (Nephrotoxins)

Ochratoxins, produced primarily by *Aspergillus ochraceus* and *Penicillium viridicatum*, cause listlessness, huddling, diarrhea, tremors, and other neural abnormalities in broiler poultry.

Zearalenone, Zearalenol, and Estrogenic Syndrome

Zearalenone and related zearalenol are almost exclusively produced by *Fusarium roseum* f. sp. *cerealis* (*F. graminearum*) growing in ears of corn standing in the field or in stored ear maize. When consumed by swine in more than a few parts per million, these compounds cause the estrogenic syndrome, externally characterized by a swollen and edematous vulva in females (Fig. 65) and enlarged mammary glands in young males. The financial loss to farmers is caused

Table IV. Mycotoxins and Toxin-Producing Fungi from Corn and Effects on Animals

Toxin or Syndrome	Fungal Source	Possible Effects on Animals
Aspergillus Toxins (primarily)		
Aflaxotins B_1, B_2, G_1, and G_2. (B_{2a}, G_{2a}, M_1, and M_2 are metabolites and are seldom present in grain; M_1 and M_2 are important contaminants in milk)	*Aspergillus flavus* and *A. parasiticus*	Hepatotoxin; carcinogenic; reduced growth rate; hemorrhagic enteritis
Ochratoxins (nephrotoxins)	*Aspergillus ochraceus* and *Penicillium viridicatum*	Toxic to kidneys and liver; abortion
Sterigmatocystin	*Aspergillus nidulans* and *A. versicolor*	Toxemia; carcinogenic
Tremorgenic toxin	*Aspergillus flavus, Penicillium cyclopium,* and *P. palitans*	Tremors and convulsions
Penicillium Toxins (primarily)		
Patulin	*Penicillium urticae, P. expansum, P. claviforme,* and *Aspergillus clavatus*	Hemorrhages of lung and brain; edema; toxic to kidneys; possibly carcinogenic
Rubratoxin	*Penicillium rubrum*	Liver damage and hemorrhaging
Citrinin	*Penicillium citrinin*	Kidney damage
Fusarium Toxins		
Emetic factor, vomitoxin (deoxynivalenol)	*Fusarium roseum* f. sp. *cerealis* (sexual stage is *Gibberella zeae*)	Vomiting; refusal by swine, cats, dogs
Feed refusal factor	May be the same as above	Feed refusal by swine
Other trichothecenes (T-2, monoacetoxyscirpenol [MAS], diacetoxyscirpenol [DAS])	*Fusarium tricinctum,* some strains of *F. roseum* f. sp. *cerealis, F. equiseti,* and *F. lateritium*	Inflamation of gastrointestinal tract and possible hemorrhage; edema; vomiting and diarrhea; infertility; degeneration of bone marrow; death
Zearalenone and zearalenol (estrogenic syndrome)	*Fusarium roseum* f. sp. *cerealis, F. tricinctum,* and in a minor way by *F. moniliforme*	Hyperestrogenism and infertility. *F. moniliforme* may cause encephalomalacia in horses

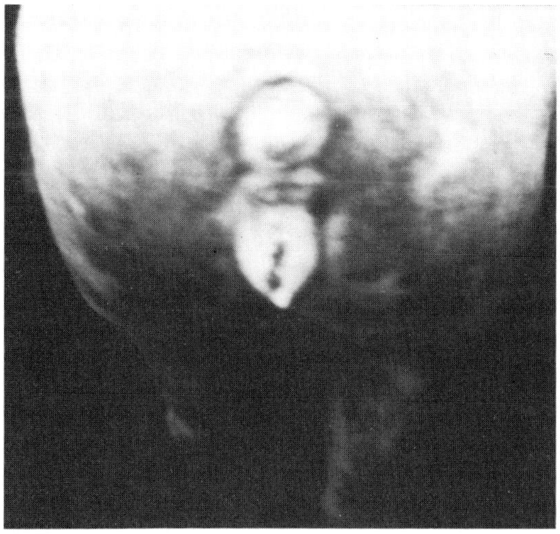

Fig. 65. Swollen and edematous vulva in sow caused by consumption of maize infected with *Fusarium roseum* f. sp. *cerealis* (*F. graminearum*). (Courtesy B. J. Jacobsen)

primarily by poor reproductive performance.

Estrogenism in swine is usually most prevalent in winter and early spring because the fungus requires a period of relatively low temperatures to produce biologically significant amounts of zearalenone or zearalenol. Some strains of *F. roseum* f. sp. *cerealis,* when growing in maize, produce a mixture of toxins along with zearalenone, and one or more of these, when consumed by swine, can cause severe stunting and other deleterious effects.

Refusal Factor and Vomitoxin (Deoxynivalenol)

Fusarium roseum f. sp. *cerealis* (sexual stage *Gibberella zeae*) (see pages 52 and 56), growing in maize ears before harvest or in storage, may produce several toxins that make the grain unpalatable to swine. Field-infected maize with more than about 5% visibly damaged kernels (see Plate 40C) is refused by pigs. The refusal toxin cannot be inactivated by any practical treatment. Growing in corn before harvest, the same fungus at times produces vomitoxin (deoxynivalenol) that causes refusal and occasional prolonged vomiting. This toxin is a trichothecene. Deoxynivalenol, T-2 toxin, monoacetoxyscirpenol, and diacetoxyscirpenol are also trichothecenes and may be involved in the refusal factor and in other cases of mycotoxicoses not yet specifically identified. Other trichothecene toxins (see below) may cause unpalatability and be involved in the refusal factor. Vomitoxin, zearalenone, and T-2 toxin have been found in stalks of corn decayed by *Fusarium*.

T-2 Toxins and Other Trichothecenes

Fusarium tricinctum and some strains of *F. roseum* f. sp. *cerealis, F. equiseti,* and *F. lateritium* produce T-2 and other toxic trichothecenes. In poultry, T-2 toxins may produce lesions at the edges of the beaks, abnormal feathering, reduced egg production, eggs with thin shells, reduced body weight gain, and mortality. T-2 toxins in amounts sufficient to cause toxicoses have been found in maize in the field, in silage, and in prepared feeds made in part from maize.

Extraction and Identification of Toxins

Evidence that a mycotoxin is responsible for illness in animals that consumed the feed requires that the toxin or toxins be isolated, purified, and quantified. Procedures have been developed for the extraction, purification, and quantification of the major mycotoxins such as zearalenone, zearalenol, T-2 and some of the other trichothecene toxins, and aflatoxins. To achieve reliable results, these procedures require considerable expertise in the performance and interpretation of the tests plus sophisticated and expensive equipment. Routine handling of contaminated grain, particularly heavily contaminated grain, may present a significant health hazard to technical personnel. Samples should therefore only be handled by trained individuals, working in appropriate facilities and within the guidelines of an acceptable safety protocol.

Sampling for Mycotoxins and Sample Preparation

An adequate sample of suspect grain must be obtained for any assay method selected. Proper sampling is essential since one aflatoxin-contaminated kernel in 1,000 kernels of grain may be a source of significant contamination. Occasionally, a biased sample may be more revealing than a truly representative one. For example, in studying stored grain or feed that shows evidence of moisture damage, heating, or "caking," a sample of damaged grain may be more appropriate than a composite one from an entire lot. A 5-kg sample is usually collected by probe or continuously from a grain stream. The sample must then be finely ground and thoroughly blended to obtain a subsample appropriate for analysis. Samples stored for subsequent analysis should be kept under cool, dry conditions that will not permit fungal growth and possible synthesis of mycotoxin(s).

Aflatoxin Detection

Methods for aflatoxin analysis fall into two categories: rapid screening procedures determining the presence or absence of aflatoxin (fluorometric iodine rapid screening and minicolumn tests), and laboratory procedures quantifying the actual amounts of toxin (thin-layer chromatography, gas-liquid chromatography, or high-pressure liquid chromatography tests). Various commercial, state, and federal laboratories perform aflatoxin analyses on a fee basis. Several confirmation tests are available for the identification of aflatoxins.

Zearalenone and Trichothecene Detection

The estrogenic syndrome, feed refusal, or vomiting may be one of the first signs of ingestion of a *Fusarium* toxin. If analysis is required, then thin-layer chromatography, gas-liquid chromatography, high-pressure liquid chromatography, or mass spectroscopy may be utilized.

Diagnosis

Few mycotoxins produce clinical signs in animals so characteristic that they permit unequivocal diagnosis. For example, the estrogenic syndrome in swine can be caused by diethylstilbestrol in the feed as well as by zearalenone and zearalenol. Refusal of corn or feed containing corn is indicative of *Fusarium* toxins; at times other contaminants may lead to refusal. Some mycotoxins, including both the trichothecenes and aflatoxins, may result in reduced productivity or depressed growth by the consuming animals. Other environmental factors may cause similar effects. Lowered animal resistance to infection by microorganisms or opportunistic parasites and reduced effectiveness of immunization may be the result of, but not limited to, the ingestion of mycotoxins.

Minimizing Mycotoxin Production after Harvest

1. Harvest as soon as the moisture content allows minimum grain damage (usually 24 to 26% moisture).
2. Adjust the harvesting equipment for minimum kernel damage and maximum cleaning.
3. Dry the shelled grain to at least 15% moisture as rapidly as possible but not longer than 24 to 48 hr after harvest. Safe, long-term storage can be achieved at a uniform moisture content of 13%. Slow drying, accomplished by low heat or natural air drying, is increasingly being used in some regions, but the grain can be *no* higher than 20–21% moisture in full bin drying. Another possibility is high temperature drying until the grain reaches 20–21% moisture followed by low heat drying to 13% moisture.
4. Cool the grain after drying and maintain dry storage conditions. Where possible, continue cooling until grain temperature reaches 2–5°C.
5. Clean the grain and bins thoroughly before storage to remove dirt, dust and other foreign matter, crop debris, chaff, and cracked or broken kernels.
6. Store in watertight structures that are free from insects and rodents.
7. Continue periodic aeration and probing for "hot spots" at intervals of one to four weeks throughout the storage period.
8. Several organic acids—propionic, isobutyric, acetic, and mixtures of these with ammonium isobutyrate—are registered for use on high-moisture grain in storage and sold under various trade names. Although these acids will not remove any aflatoxins previously formed in the grain, they will prevent growth of fungi if properly applied. *Grain treated with an organic acid can be used only for livestock and poultry feed.*

Selected References

PURCHASE, I. F. H. (ed.). 1974. Mycotoxins. Elsevier: Amsterdam.

RODRICKS, J. V. (ed.). 1976. Mycotoxins and other fungal related food problems. Adv. Chem. Ser. 149. Am. Chem. Soc.: Washington, DC.

RODRICKS, J. V., C. W. HESSELTINE, and M. A. MEHLMAN (eds.). 1977. Mycotoxins in Human and Animal Health. Pathotox Publishers: Park Forest South, IL.

SHOTWELL, O. L. 1977. Aflatoxin in corn. J. Am. Oil Chem. Soc. 54:216A-224A.

WYLLIE, T. D., and L. G. MOREHOUSE (eds.). 1977-1978. Mycotoxic Fungi, Mycotoxins, Mycotoxicoses. An Encyclopedic Handbook. 3 vols. Dekker: New York.

Viral Diseases

The more than 600 plant-infecting viruses are macromolecules composed of either ribonucleic acid (RNA) or deoxyribonucleic acid (DNA) surrounded by a protective protein or lipoprotein coat. Virus particles, or virions, can only replicate within living cells and thus are obligate parasites. The virions can be filamentous (long rigid or flexuous rods), roughly spherical (isometric), or bacilliform in shape. They range in diameter from 10 to 70 nm, and rod lengths can exceed 2 μm. All those that infect maize, except maize streak virus, contain RNA. Observation of virus particles requires a high-resolution electron microscope (Fig. 3).

Viruses are transmitted to plants through pollination and through wounds created by animal (mainly arthropod and nematode) or fungal vectors, parasitic plants, mechanical inoculation, and man's deliberate or accidental activities (e.g., propagation of plants by cuttings, grafts, or budding and by planting infected seed).

Viruses are identified by: 1) host specificity, 2) particle morphology, 3) mode(s) of transmission, and 4) their biochemical, physical, and serological properties. The specific reactions of indicator plant hosts also aid in identification. Most plant viruses are named for the original or major host attacked, and the symptoms they produce. Plant viruses are taxonomically grouped by shape and size, mode of transmission, serology, and antigenic properties.

Viral infections range from latent to premature plant death. They may induce stunting, excessive tillering, mosaic patterns, yellowing or reddening of the foliage, and necrosis. The action of two or more viruses in a given plant (multiple infections) can be additive, synergistic, or cross-protective. Virus-infected plants are often more susceptible to root or stem (stalk) rots and possibly other types of diseases.

More than 40 viruses or virus strains have been reported to cause diseases of maize throughout the world

(Tables V and VI). Several of these diseases are not known to occur in nature.

At present, identification of virus and viruslike diseases *in the field* is more an art or "game of chance" than a science. Satisfactory clinical techniques, however, are available to identify many viruses. Indistinct symptomatology, the occurrence of different viruses in the same area, the existence of virus strains, the presence of singly and doubly infected plants, and inadequate characterizations are a few of the factors that confuse disease identification in the field. In addition, symptoms of virus infection can be confused with those caused by mycoplasma-induced diseases (pages 11–12). Genetic and nutritional abnormalities may be ruled out by transmission of a causal agent from diseased to healthy plants, electron microscopy of diseased tissue to observe virus particles that are absent in healthy plants, and serology to detect relationships with known plant viruses.

Sugarcane Mosaic

Several closely related viral diseases distributed worldwide are caused by strains and substrains of sugarcane mosaic virus (SCMV) (Table V). Innumerable variants or substrains of SCMV can be identified by extensive host range studies, but this soon becomes impractical. The most important differential hosts of SCMV are its perennial reservoirs, most frequently, sugarcane, or Johnsongrass.

SCMV infection of maize (Figs. 66, 67, and Plate 48) has occurred in most tropical and subtropical regions where maize was grown next to infected sugarcane. This disease problem has now largely been solved by use of resistant sugarcane varieties. The important problem remaining is the introduction and spread in temperate climates of virus strains that survive in perennial grasses, e.g., the survival of some maize dwarf mosaic virus strains in Johnsongrass.

Maize Dwarf Mosaic

Symptoms are highly variable and the following description depicts frequently seen effects. In the bases of the youngest leaves, irregular, light and dark green mottle or mosaic patterns appear that may develop into narrow, light green or yellowish streaks along the veins. The mosaic often appears as dark green "islands" on a chlorotic background (Fig. 66). Maize dwarf mosaic symptoms may be present on all leaves, leaf sheaths, and ear leaves (husks) that develop following infection. Plants are sometimes slightly stunted and reduction in ear size and seed set may occur. As infected plants continue to grow and the ambient temperature rises, the mosaic symptoms often disappear and young leaves become more yellow. Early infection may predispose maize to root and stalk rots and premature death. Symptoms have been observed in the field within 15 days after seedling emergence.

MDMV is commonly confused with maize stunt and maize bushy stunt (pages 11–12).

Maize dwarf mosaic virus (MDMV) is a long, flexuous, rod-shaped virion 12–15 × 750 nm (Fig. 3) that is transmitted mechanically. The virus is retained and transmitted in a nonpersistent manner (up to 6 hr) by over 20 species of aphids including the corn leaf aphid,

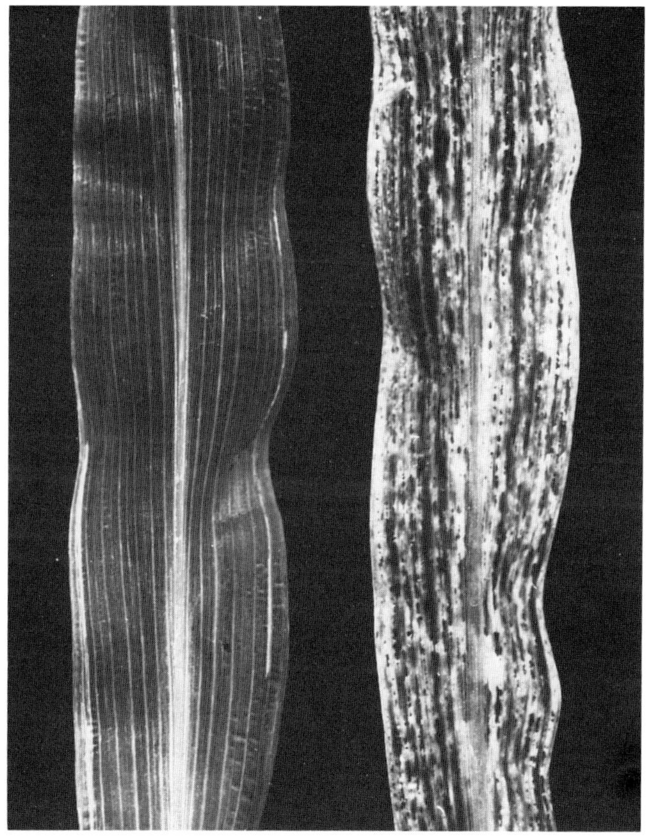

Fig. 66. Sugarcane mosaic. Maize leaf at right shows typical chlorotic spots and streaks due to SCMV; healthy leaf at left. (Courtesy M. B. von Wechmar)

Table V. Strains and Substrains of Sugarcane Mosaic Virus (SCMV) that Cause Disease in Maize

Disease (Virus)	Perennial Grass Host	Geographical Distribution
Sugarcane mosaic	Sugarcane[a]	Worldwide in tropics and subtropics
Maize dwarf mosaic, strains A,C,D,E,F	Johnsongrass	Worldwide
Maize dwarf mosaic, strain B[b]	...	Worldwide
Maize mosaic[c]	...[d]	India
Maize mosaic[c] (Abaca mosaic)	Johnsongrass	Philippines
Sorghum concentric ring blotch	...[d]	South Africa
Sorghum red stripe	Johnsongrass[d]	France to the USSR
Transvaal grass mosaic	*Sorghum arundinaceum*[d]	South Africa

[a] At least 13 strains of SCMV, A through M, are known in the USA and are identified by use of sugarcane varieties.
[b] Any SCMV strain that infects corn and sugarcane but not Johnsongrass would presently be considered maize dwarf mosaic virus, strain B.
[c] Not to be confused with maize mosaic caused by a rhabdovirus.
[d] Natural occurrence in commercial sugarcane has not been reported.

Table VI. Other Viral Diseases Occasionally Reported on Maize[a]

Disease	Other Hosts	Primary Symptoms	Vectors-Transmission	Distribution
Barley stripe mosaic	Barley, oats, wheat, and numerous other grasses	Leaf striping	Transmitted by seed	Worldwide
Barley yellow dwarf	Barley, oats, wheat, and numerous other grasses	Chlorosis of leaf tips and margins; intensification of underlying colors (dark red to purple, orange), older leaves first. Stunting is slight	*Rhopalosiphum padi, R. maidis, Schizaphis graminum, Sitobion avenae, S. fragariae, S. miscanthi,* and seven other aphids (persistent)	Widespread in other hosts in North America, Europe, Asia, Africa, Australia, and New Zealand
Brome mosaic	Bromegrass, wheat, oats, bermudagrass, wheatgrasses, orchardgrass, beardgrasses or bluestems, numerous other perennial grasses, and some dicotyledons	Fine or broad chlorotic stripes in young leaves. Seedlings or older plants may wilt and die	Transmission by maize rootworms or beetles observed in the USSR and USA (*Diabrotica undecimpuncata, D. virgifera*);nematodes reported in Eastern Europe (*Longidorus macrosoma, Xiphinema coxi, X. diversicaudatum*)	Probably worldwide in other hosts
Cereal chlorotic mottle	Barley, oats, wheat, sweetcorn, several grasses	Initial chlorotic stripes followed by mild chlorotic mottle	*Nesoclutha pallida*	Queensland, Australia
Cereal tillering disease	Barley, oats, grasses	Shoot proliferation, stunting, vein enations	*Laodelphax striatellus, Dicranotropis hamata* (persistent)	Sweden
Cucumber mosaic	Wheat, rye, many monocotyledons and dicotyledons	Chlorotic, elliptical spots that form leaf stripes (Fig. 73); if severe, plants are stunted with necrotic leaf spots or death of seedlings	*Aphis gossypii, Myzus persicae,* and many other aphid species (nonpersistent and stylet-borne)	Probably worldwide in other hosts
Fiji disease	Sugarcane, sorghum	Galls on veins	*Perkinseilla saccharicida, P. vastatrix* (persistent)	Philippines, New Guinea, Java, New South Wales
Maize hoja blanca[b]	*Rottboellia exaltata*	Narrow-to-wide, whitish to yellow stripes; new leaves are almost white; slight stunting	*Peregrinus maidis* (persistent)	Venezuela
Maize leaf fleck	*Phalaris stenoptera* (Harding grass)	Small, round, pale yellow to orange leaf spots; marginal burn; leaf tips and older leaves are affected first	*Myzus persicae, Rhopalosiphum padi, R. maidis* (persistent)	San Francisco Bay area (California USA)
Maize line	...	Broad yellow bands on leaves	*Peregrinus maidis* (persistent)	East Africa
Maize pellucid ringspot	...	Round, water-soaked leaf spots	An aphid is suspected	West Africa, Papua
Maize red stripe	...	Red streaks on leaves; dwarfing; small cobs	Unknown	Bulgaria
Maize ring mottle	...	Chlorotic spots and rings on leaves	Transmitted by seed	Bulgaria
Maize tassel abortion	...	Small leaves horizontal to stem; tassel abortion	*Malaxodes farinosus*	East Africa
Maize vein enation	Wheat, rice, rye, sorghum, sugarcane, oats, grasses	Galls on lower leaf surfaces; dark foliage; stunting	*Cicadulina mbila* (persistent)	India
Millet red leaf	*Panicum* and *Setaria* species	Reddening of leaf blades and sheaths of purple-stem varieties; yellowing of green-stem varieties	*Macrosiphum granarium, Rhopalosiphum maidis, Schizaphis graminum*	China
Northern cereal mosaic	Barley, oats, wheat, rye, and several grasses	Chlorotic striping along one side of leaf blades	*Laodelphax striatellus, Delphacodes albifascia, Unkanodes sapporonus, Meulerianella fairmairei*	Japan, Korea
Oat pseudorosette (Zakuklivanie)	Oats, barley, rye, rice, wheat, *Panicum miliaceum, Echinochloa crusgalli, Agropyron repens, Bromus* and *Setaria* species	Reddish-brown discoloration; excessive tillering; inflorescence malformation	*Laodelphax striatellus* (persistent)	Western Siberia
Oat sterile dwarf	Barley, oats, grasses	Ephemeral enations; narrow leaves	*Javesella pellucida, J. discolor, J. obscurella, Dicranotropis hamata* (persistent)	Sweden
Rice black-streaked dwarf	Rice, barley, wheat, and many grasses	Chlorotic leaf streaks; stunting of plants	*Laodelphax striatellus, Unkanodes sapporonus, Ribautodelphax (U.) albifascia* (persistent)	Japan
Rice stripe	Rice, barley, wheat, and millets	Chlorotic leaf streaks; stunting of plants	*Laodelphax striatellus, Unkanodes sapporonus, Ribautodelphax albifascia* (persistent)	Japan
Tsai's disease[b]	Sweet corn	Numerous narrow chlorotic stripes	*Peregrinus maidis* (persistent)	Florida (USA)
Wheat spot mosaic	Wheat, barley, rye, and other grasses	Chlorotic leaf spots, streaks, and mottling	*Eriophyes tulipae* (persistent)	Canada, Ohio (USA)

[a] Reports of infection on maize in nature are missing for some viruses.
[b] Shown to be caused by a serologically related entity believed to be the same as the maize stripe virus.

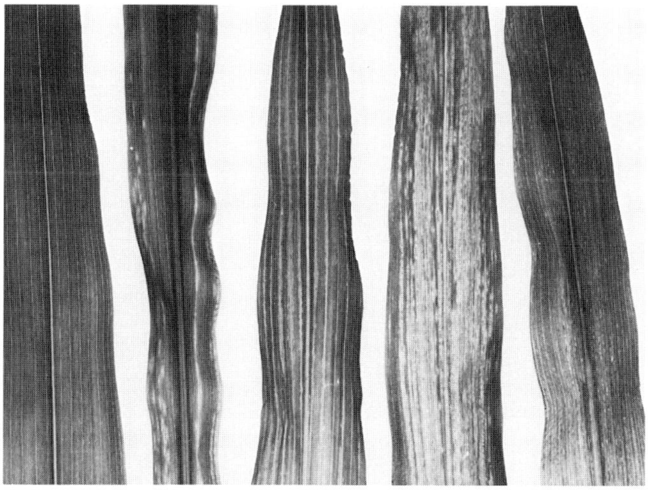

Fig. 67. Maize dwarf mosaic. Typical range of mottling and streaking symptoms produced by MDMV on maize leaves; healthy leaf at left. (Courtesy R. E. Ford)

Rhopalosiphum maidis (Fitch), the greenbug, *Schizaphis graminum* (Rondani), and the green peach aphid, *Myzus persicae* (Sulzer). Over 200 wild and cultivated grasses are susceptible to MDMV, but Johnsongrass is believed the principal overwintering reservoir host for strains that infect it (Table V). However, outbreaks of MDMV-A in late-planted dent and sweet corn have occurred in areas up to 400 miles north of where Johnsongrass overwinters as rhizomes. Vectors acquire MDMV-A from Johnsongrass in the spring and summer and transmit it to maize plants. Strain B does not infect Johnsongrass and is apparently introduced into maize plantings from another source(s). MDMV is seedborne in sweet maize (up to 0.05%) and possibly certain annual grassy weeds, and this may account in part for the widespread distribution of maize dwarf mosaic.

Maize Chlorotic Dwarf

The earliest symptom caused by maize chlorotic dwarf virus (MCDV) is a chlorosis of the young leaves in the whorl. As infected leaves unfurl, a distinct chlorotic striping bordering the secondary and tertiary veins becomes apparent. Leaf yellowing or reddening and moderate to severe stunting of internodes are common, the degree dependent on the susceptibility of the hybrid or variety (Plate 49).

Double infections of MCDV and MDMV occur in the eastern USA. These double or mixed infections probably cause a more serious disease (chlorosis and severe stunting or dwarfing) than single infections.

MCDV is transmitted by the leafhopper *Graminella nigrifrons* (Forbes) and *G.* (*Deltocephalus*) *sonora* (Ball) in a semipersistent manner. After acquiring the virus by feeding on infected plants, nymphs and adults are viruliferous for up to 48 hr.

The isometric (31 nm in diameter) maize chlorotic dwarf virion contains RNA. The main overwintering host is Johnsongrass. The distribution of MCDV is from the southern edge of the Corn Belt to the southeastern USA and coincides with the overlapping distributions of *G. nigrifrons* and Johnsongrass. MCDV has not been reported outside of the USA. Other hosts include broomcorn millet, milo, grain sorghum, Sudangrass, wheat, crabgrass, and foxtails (*Setaria* spp.). Tolerant but not immune maize varieties have been developed. Herbicides to eliminate Johnsongrass and systemic insecticide sprays to reduce leafhopper populations also result in disease control.

Maize Chlorotic Mottle

First described in 1974 as a new disease in Peru, maize chlorotic mottle virus (MCMV) produces fine chlorotic streaks in the youngest leaves 7–10 days after inoculation. These streaks coalesce into uniform, systemic chlorotic mottling followed by necrosis, epinasty, and death of leaves. Growth is stunted, male inflorescences are distorted, and fewer ears are formed (Plate 50). The virus particles are isometric and measure about 30 nm in diameter.

The disease first appeared in the USA in 1976, and has been found in areas of Kansas and Nebraska. The virus can be transmitted mechanically and by six economically important leaf beetles (Chrysomelids): *Oulema melanopa* L. (cereal leaf beetle), *Chaetocnema pulicaria* Melsheimer (corn flea beetle), *Systema frontalis* (Fab.) (flea beetle), *Diabrotica undecimpunctata* Mannerheim (southern corn rootworm), *D. longicornis* (Say.) (northern corn rootworm), and *D. virgifera* LeConte (western corn rootworm). Other hosts of MCMV include wheat, barley, rye, sorghum, Johnsongrass, teosinte, and grasses (*Andropogon, Bromus, Digitaria, Eragrostis, Panicum, Setaria,* and *Spartina* species).

Infections by MCMV cause a mild disease in dent and flint maize types (floury and sweet are more susceptible) but mixed infections with maize dwarf mosaic or wheat streak mosaic result in a severe disease described as corn lethal necrosis (see below).

Maize Mosaic

Symptoms of this disease (also called "enanismo rayado," corn stripe, and sweet corn mosaic) appear as chlorotic spots, short lines, and broken to nearly continuous fine-to-broad chlorotic stripes centered on the secondary or tertiary veins (Fig. 68). Leaves commonly have long, broad, yellow stripes that finally become necrotic. Chlorotic stripes can also occur on the leaf sheath, ear husks, and stalks. Moderate to severe dwarfing or rosetting (brachyism) of new growth can occur, depending on the plant's age when infected and the genetic variability of the hybrid or variety. Stalk size, leaf blades, and tassel tend to be normal. Young plants are most susceptible.

Maize mosaic virus (MMV) is transmitted in a persistent manner by the corn planthopper, *Peregrinus maidis* (Ashmead). The particle is a rhabdovirus that measures 48 × 242 nm in plant sections and 90 × 225 nm in leaf dips. Plant rhabdoviruses multiply in their insect vectors during a long incubation period before the insects are able to transmit the virus to plants during the remainder of their lives. The incubation period in maize seedlings ranges from 4 to 30 days. The disease also occurs in sorghum and a few wild grasses, e.g., *Rottoellia exaltata, Setaria vulpiseta,* and *Zea mays mexicana*. MMV occurs in the tropics and neotropics, specifically in Hawaii, Cuba, Puerto Rico, Trinidad, northern South America, West Africa, Madagascar,

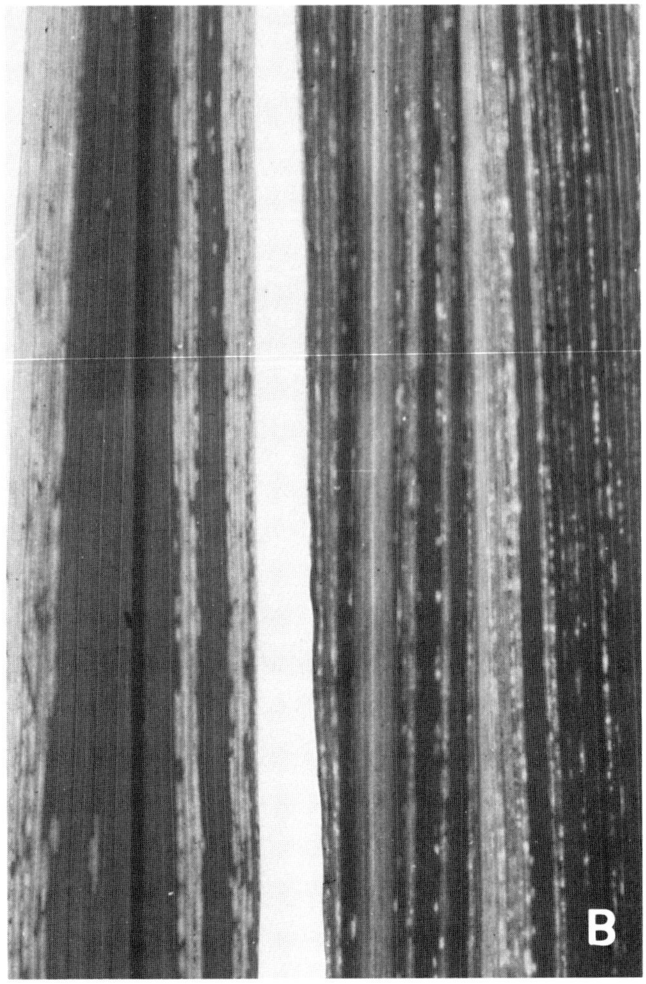

Mauritius, India, Australia (Tasmania), and in the USA (Alabama, Iowa, Louisiana, Mississippi, Nebraska, and Texas).

When maize mosaic and maize streak (see below) occur in the same locality, the two diseases can be confused because of similarity of symptoms. In these cases, other tests or transmission by insect vectors is needed to identify the two viruses.

Maize Rayado Fino or Fine Striping Disease

Early symptoms consist of chlorotic dots, spots, or short lines distributed more or less uniformly over the youngest leaf and centered on the secondary and tertiary veins. The dots gradually become more numerous and may fuse longitudinally forming chlorotic stripes up to 10 cm long. These symptoms are more pronounced in seedlings. Occasionally, stunting

←

Fig. 68. Maize mosaic, sometimes called corn stripe. **A,** Close-up of leaf showing typical chlorotic spots, short lines, and broken to nearly continuous chlorotic stripes. (Courtesy Frieda Herold) **B,** Close-up of maize leaves illustrating streaks and stripes centered on leaf veins and broad bands formed by fusing of the stripes. **C,** Field plant infected at a later stage than B. (Courtesy Queensland Department of Primary Industries)

and chlorosis are observed (Plate 51). Plants infected at tasseling or later do not develop symptoms. Yields may be reduced up to 45–50% in susceptible hybrids and varieties.

Maize rayado fino virus (MRFV) is transmitted in a persistent manner by the leafhopper *Dalbulus maidis* (Delong & Wolcott) (Fig. 13), *D. elimatus* (Ball), *Graminella nigrifrons* (Forbes), *Baldulus tripsaci* Kramer and Whitcomb, and *Stirellus bicolor* (Van Duzee), following a latent period of 7 to 22 days. Insects retain infectivity and transmit intermittently up to four weeks. The virus particles are isometric or polyhedral-shaped, 30 nm in diameter, and restricted to the phloem of infected plants. Maize rayado fino is widespread in maize-growing regions of South and Central America and has been found in the southern USA (Florida and Texas). This virus has been misinterpreted as a variant of corn or maize stunt (page 11) since it is transmitted by the same insects. Several annual and perennial species of *Zea* and *Tripsacum* as well as *Rottboellia exaltata* are hosts of MRFV.

Fig. 69. Wheat streak mosaic. Early symptoms showing chlorotic spots and broken streaks on a young leaf of Golden Bantam sweet maize. (Courtesy A. M. Finley)

Maize Colombian stripe (del rayado Columbiana del maiz) and Brazilian corn streak are diseases of maize caused by strains of MRFV.

Maize Streak

Initial symptoms consist of minute, round, cream-to-white scattered spots on the youngest leaves (Plate 52). These spots later elongate and coalesce into long, discontinuous chlorotic streaks distributed uniformly over all leaf surfaces formed later (Plate 53). The parallel chlorotic streaks may fuse partially or completely, leaving irregular green lines or green islands centered between the veins. The streaks are most conspicuous in the youngest leaves. Fully formed leaves develop a chlorosis with broken yellow streaks along the veins that contrast with the dark green of normal leaves. Early infection of highly susceptible plants can drastically reduce stalk internodes and leaf size. There may be no stunting of plants infected late in the season. Early infection may result in ears that are poorly filled or that lack viable kernels.

Maize streak virus (MSV) is transmitted by five leafhopper species of the genus *Cicadulina* of which *C. mbila* (Naude) is the most important. The maize streak virions are isometric, 20 nm in diameter, and frequently occur in pairs measuring 20 × 30 nm. The nucleic acid of MSV is a circular, single-stranded DNA. Various strains of MSV are known, each specialized to a particular host species. MSV infects wheat, barley, oats, rye, rice, sugarcane, teosinte, broomcorn millet, African millet, and many wild grasses. It is widespread in Africa south of the Sahara Desert, India, and has been reported from the islands of Mauritius, Madagascar, and Reunion. Bajra streak (India) and maize mottle (Africa) are caused by strains of MSV.

Maize Stripe

The first symptoms of maize stripe, caused by the maize stripe virus (MStpV), are numerous chlorotic spots and narrow streaks that appear at the base and extend outward on the youngest leaves. As these leaves expand, chlorotic spots and stripes fuse to form broad chlorotic bands that extend to the leaf bases. Chlorotic patterns involve the entire leaf blade of later-developing

Fig. 70. Wheat curl mites feeding on a maize leaf. (Courtesy H. H. Thornberry)

leaves. A tendency for acute bending at the apical tip, stunting, and reduced yields in severely diseased plants may also occur.

MStpV is transmitted in a persistent manner by the corn planthopper, *Peregrinus maidis*. A long, rod-shaped particle is associated with maize stripe disease. It is serologically related or identical to entities causing maize hoja blanca and Tsai's disease (Table VI). The host range of MStpV is limited to the close relatives of maize (teosinte species), sorghum, barley, and the annual weed *Rottboellia exaltata*. The virus originally was described from East Africa (as an isometric virus) and has since been identified from West Africa, Venezuela (where it is called "maize hoja blanca"), Peru, and southern Florida in the USA (Tsai's disease). The virus probably occurs in all tropical regions worldwide where the *P. maidis* vector occurs.

Wheat Streak Mosaic

Early symptoms appear as small chlorotic spots or broken streaks at the *tips* of young leaves (Fig. 69). The

→

Fig. 71. Maize rough dwarf. **Upper left,** vein galls on lower leaf surfaces. (Courtesy I. Harpaz) **Upper right,** severely reduced root system showing shortness of brace roots. (Courtesy P. Grancini). **Lower right,** MRDV-induced atrophy of maize ears compared with noninfected ear that suffered from deficient pollination (left). (Courtesy P. Grancini)

streaks elongate and develop parallel to the veins. Older leaves may become chlorotic near the tips with green margins bordering the veins. Symptoms vary greatly with the plant genotype. Very susceptible lines may have poorly developed ears with little or no seed set, and general yellowing and stunting of the plant may occur. Corn inbreds are particularly susceptible when young.

Transmission of wheat streak mosaic virus (WSMV) is by the minute wheat-curl mite, *Eriophyes (Aceria) tulipae* (Keifer) (Fig. 70). WSMV infects wheat, other cereals, and a wide range of annual and perennial grasses in North America, Jordan, USSR, Romania, and Yugoslavia. WSMV virions are long flexuous rods of 15×700 nm.

Maize Rough Dwarf

Maize rough dwarf virus (MRDV) dwarfs all parts of infected plants, inducing small swellings (galls) on the veins on the lower surfaces of young leaves. General chlorosis or chlorotic streaks may occur on the new growth. Reddish discoloration may begin at the leaf margins and eventually involve the entire blade. Root systems of infected plants are fragile, discolored, and greatly reduced. Ears on affected plants are dwarfed and poorly filled. Symptoms appear 15–20 days after inoculation (Fig. 71). MRDV is a double-stranded RNA phytoreovirus. The polyhedral virions, about 70 nm diameter, are transmitted by the planthoppers *Laodelphax striatellus* (Fallen), *Javesella pellucida* (Fab.), *Delphacodes propinqua* Fieber, and *Sogatella vibix* (Haupt.). Only *L. striatellus* is a known vector in nature. Phytoreoviruses multiply in their insect vectors during a long latent period before being able to transmit the virus to plants for the remainder of their lives.

Maize rough dwarf (also called "nanismo ruvido") infects maize, wheat, barley, oats, bermudagrass, crabgrass, and barnyard grass in southern Europe, Israel, and India.

Maize Wallaby Ear

Symptoms of maize wallaby ear disease occur in two forms: mild consisting of ephemeral galls and enations followed by plant recovery, and severe with numerous galls on most or all secondary veins, severe stunting, and no grain yield.

Severely affected plants have upright dark green leaves with edges rolled upward and inward that stand out stiffly at right angles to the stalk (Fig. 72). Maize wallaby ear virus (MWEV) is a double-stranded RNA phytoreovirus that is transovarially passed to vector offspring. Three leafhopper vectors are *Cicadulina bipunctella bimaculata* Evans, *C. bimaculata* (Evans), and *Nesoclutha pallida* (Evans). MWEV is distantly related to maize rough dwarf virus, Fiji disease virus,

Fig. 72. Wallaby ear. Maize plants showing the difference in appearance of healthy and diseased plants. (Courtesy Queensland Department of Primary Industries)

Fig. 73. Cucumber mosaic. Leaves of Golden Cross Bantam sweet corn, illustrating mottle **(right)** and wide bands of chlorotic tissue **(left).** (Courtesy N. W. Frazier and J. H. Freitag)

and rice black-streaked dwarf virus (Table VI). It has been reported only from several areas in Australia, although the vectors are found in many areas including Africa and Southeast Asia. Other hosts of MWEV include rice, rye, sorghum, wheat, sugarcane, and several weed grasses.

Corn Lethal Necrosis

First described in 1976, this disease, only reported from Kansas and Nebraska, USA, has the potential to become widely spread. Yield losses in severely infected fields have been estimated at 50% or more. Yields in highly susceptible hybrids were reduced 70% when mechanically inoculated at the 10–12 leaf stage.

Corn lethal necrosis results from the synergistic interaction of maize chlorotic mottle virus (MCMV) in combination with different strains of either maize dwarf mosaic (MDMV-A and -B) or wheat streak mosaic virus (WSMV). Symptoms include severe yellow mottling followed by browning of the leaves and early death of plants (Plate 54).

Leaf necrosis progresses inward from the margins. In maturing plants, necrosis usually begins at the tassel and progresses downward. Ears frequently are not produced on affected plants; those that form are reduced in size, distorted, poorly filled, and often severely affected by ear rots. The seeds produced are often wrinkled and shriveled.

Vectors include various beetle species for MCMV, aphid species for MDMV, and the wheat-curl mite for WSMV. Disease tolerance has been identified in some maize inbreds and hybrids. Crop rotation may be an effective control measure.

Control Recommendations of Various Viral Diseases

1. Rotate crops excluding grasses and cereals.
2. Use herbicides to eliminate Johnsongrass and other overwintering hosts.
3. Apply timely insecticide applications to reduce vector populations. Oil sprays may reduce virus spread in a nonpersistent manner by aphids.
4. Plant resistant or tolerant lines (inbreds), hybrids, and varieties where viral diseases are troublesome. For example, inbreds resistant to strain A of MDMV may be susceptible to strain B. All current maize inbreds resistant to strain B are also resistant to strain A.
5. Plant early to avoid insect transmission of certain viruses, e.g., MDMV.

Selected References

BOCK, K. R. 1974. Maize streak virus. Commonw. Mycol. Inst. Descr. Plant Viruses 133.

BRADFUTE, O. E., L. R. NAULT, D. T. GORDON, D. C. ROBERTSON, R. W. TOLER, and C. W. BOOTHROYD. 1980. Identification of maize rayado fino virus in the United States. Plant Dis. 64:50-53.

BRAKKE, M. K. 1971. Wheat streak mosaic virus. Commonw. Mycol. Inst. Descr. Plant Viruses 48.

CASTILLO, J., and T. T. HERBERT. 1974. Nueva enfermedad virosa afectando al maiz en al Peru. Fitopatologia 9:79-84.

GAMEZ, R. 1969. A new leafhopper-borne virus of corn in Central America. Plant Dis. Rep. 53:929-932.

GEROLA, F. M., M. BASSI, O. LOVISOLO, and C. VIDANO. 1966. Virus-like particles in both maize plants infected with maize rough dwarf virus and the vector *Laodelphax striatellus* Fallén. Phytopathol. Z. 56:97-99.

GINGERY, R. E., L. R. NAULT, J. H. TSAI, and R. J. LASTRA. 1979. Occurrence of maize stripe virus in the United States and Venezuela. Plant Dis. Rep. 63:341-343.

GORDON, D. T., O. E. BRADFUTE, R. E. GINGERY, J. K. KNOKE, and L. R. NAULT. 1979. Maize virus disease complexes in the United States: Real and potential disease problems, pp. 102-133. Proc. 33rd Annu. Corn and Sorghum Res. Conf. Am. Seed Trade Assoc.: Washington, DC.

GORDON, D. T., and L. R. NAULT. 1977. Involvement of maize chlorotic dwarf virus and other agents in stunting diseases of *Zea mays* in the United States. Phytopathology 67:27-36.

GORTER, G. J. M. A., and P. J. KLESSER. 1964. Sorghum concentric ring blotch, a newly observed virus disease. S. Afr. J. Agric. Sci. 7:329-334.

GRYLLS, N. E. 1975. Leafhopper transmission of a virus causing maize wallaby ear disease. Ann. Appl. Biol. 79:283-296.

HARPAZ, I. 1972. Maize Rough Dwarf, A Planthopper Virus Disease Affecting Maize, Rice, Small Grains and Grasses. Israel Universities Press: Jerusalem.

HEROLD, F. 1972. Maize mosaic virus. Commonw. Mycol. Inst. Descr. Plant Viruses 94.

HUTCHINSON, P. B., G. N. R. FORTEATH, and A. W. OSBORN. 1972. Corn, sorghum and Fiji disease. Sugarcane Pathol. Newsl. 9:12-13.

KITAJIMA, E. W., R. GAMEZ, and M. T. LIN. 1975. A serological and histological comparison of the maize "rayado fino" virus from Costa Rica, and the Brazilian corn streak virus. Proc. Am. Phytopathol. Soc. 2:76.

KULKARNI, H. Y. 1973. Comparison and characterization of maize stripe and maize line viruses. Ann. Appl. Biol. 75:205-216.

LINDSTEN, K., B. GERHARDSON, and J. PETTERSSON. 1973. Cereal tillering disease in Sweden and some comparisons with oat sterile dwarf and maize rough dwarf. Medd. Statens Vaxtskyddsanstalt 15:151,375-397.

LOVISOLO, O. 1971. Maize rough dwarf virus. Commonw. Mycol. Inst. Descr. Plant Viruses 72.

MARTÍNEZ LÓPEZ, G., L. M. R. DE CUJIA, and C. S. DE LUQUE. 1974. Una nueva enfermedad del maíz en Colombia transmitida por el saltahojas *Dalbulus maidis* (DeLong & Wolcott). Fitopatologia 9:93-99.

McCLEAN, A. P. D. 1947. Some forms of streak virus occurring in maize, sugarcane and wild grasses. Union S. Afr. Sci. Bull. 265.

MILNE, R. G., and O. LOVISOLO. 1977. Maize rough dwarf and related viruses. Adv. Virus Res. 21:267-341.

NAULT, L. R., D. T. GORDON, R. E. GINGERY, O. E. BRADFUTE, and J. CASTILLO LOAYZA. 1979. Identification of maize viruses and mollicutes and their potential insect vectors in Peru. Phytopathology 69:824-828.

NAULT, L. R., W. E. STYER, M. E. COFFEY, D. T. GORDON, L. S. NEGI, and C. L. NIBLETT. 1978. Transmission of maize chlorotic mottle virus by Chrysomelid beetles. Phytopathology 68:1071-1074.

NAULT, L. R., W. E. STYER, J. K. KNOKE, and H. N. PITRE. 1973. Semi-persistent transmission of leafhopper-borne maize chlorotic dwarf virus. J. Econ. Entomol. 66:1271-1273.

NIBLETT, C. L., and L. E. CLAFLIN. 1978. Corn lethal necrosis—a new virus disease of corn in Kansas. Plant Dis. Rep. 62:15-19.

ROCHOW, W. F. 1970. Barley yellow dwarf virus. Commonw. Mycol. Inst. Descr. Plant Viruses 32.

SETH, M. L., S. P. RAYCHAUDHURI, and D. V. SINGH. 1972. Bajra (pearl millet) streak: A leafhopper-borne cereal virus in India. Plant Dis. Rep. 56:424-428.

SHIKATA, E. 1974. Rice black-streaked dwarf virus.

Commonw. Mycol. Inst. Descr. Plant Viruses 135.
STOREY, H. H. 1937. A new virus of maize transmitted by *Cicadulina* spp. Ann. Appl. Biol. 24:87-94.
TRUJILLO, G. E., J. M. ACOSTA, and A. PIÑERO. 1974. A new corn virus disease found in Venezuela. Plant Dis. Rep. 58:122-126.
UYEMOTO, J. K., D. L. BOCKELMAN, and L. E. CLAFLIN. 1980. Severe outbreak of corn lethal necrosis disease in Kansas. Plant Dis. 64:99-100.
WILLIAMS. L. E., D. T. GORDON, and L. R. NAULT (eds.). 1976. Proc. Int. Maize Virus Dis. Colloq. Workshop, Wooster, OH.

Nematode Diseases

Nematodes, also called nemas or eelworms, are unsegmented roundworms that inhabit fresh and salt water, decaying organic matter, soil, plants, and animals throughout the world. They are probably the most numerous multicellular animals on earth. A pint of soil may contain 20,000 or more nematodes. About three billion nematodes have been estimated to live in an average acre of soil, most of which are in the top 6 in. Of the more than 15,000 species of nematodes described, many live in nonparasitic associations with other organisms. About 2,200 species are parasites of plants.

Plant-parasitic nematodes are typically microscopic, transparent, mobile, and vermiform (Fig. 4). Exceptions are several rather sluggish ectoparasites (e.g., *Macroposthonia* and *Criconemoides*) and the adult females of certain genera (e.g., *Heterodera* and *Meloidogyne*) whose bodies become swollen and saclike. Nematodes have bodies differentiated for feeding, digestion, locomotion, and reproduction. Some of the anatomical features used for identification include the size and shape of the adult female, the size and shape of the stylet and tail, the shape of the esophagus and reproductive organs, and cuticular patterns (Fig. 4).

Passive movement of nematodes occurs in water, soil, and infected plant parts. Nematodes are disseminated by the wind, many types of animals, tools, and vehicles and machinery carrying infested soil. Active movement of nematodes through soil is very limited, considered to

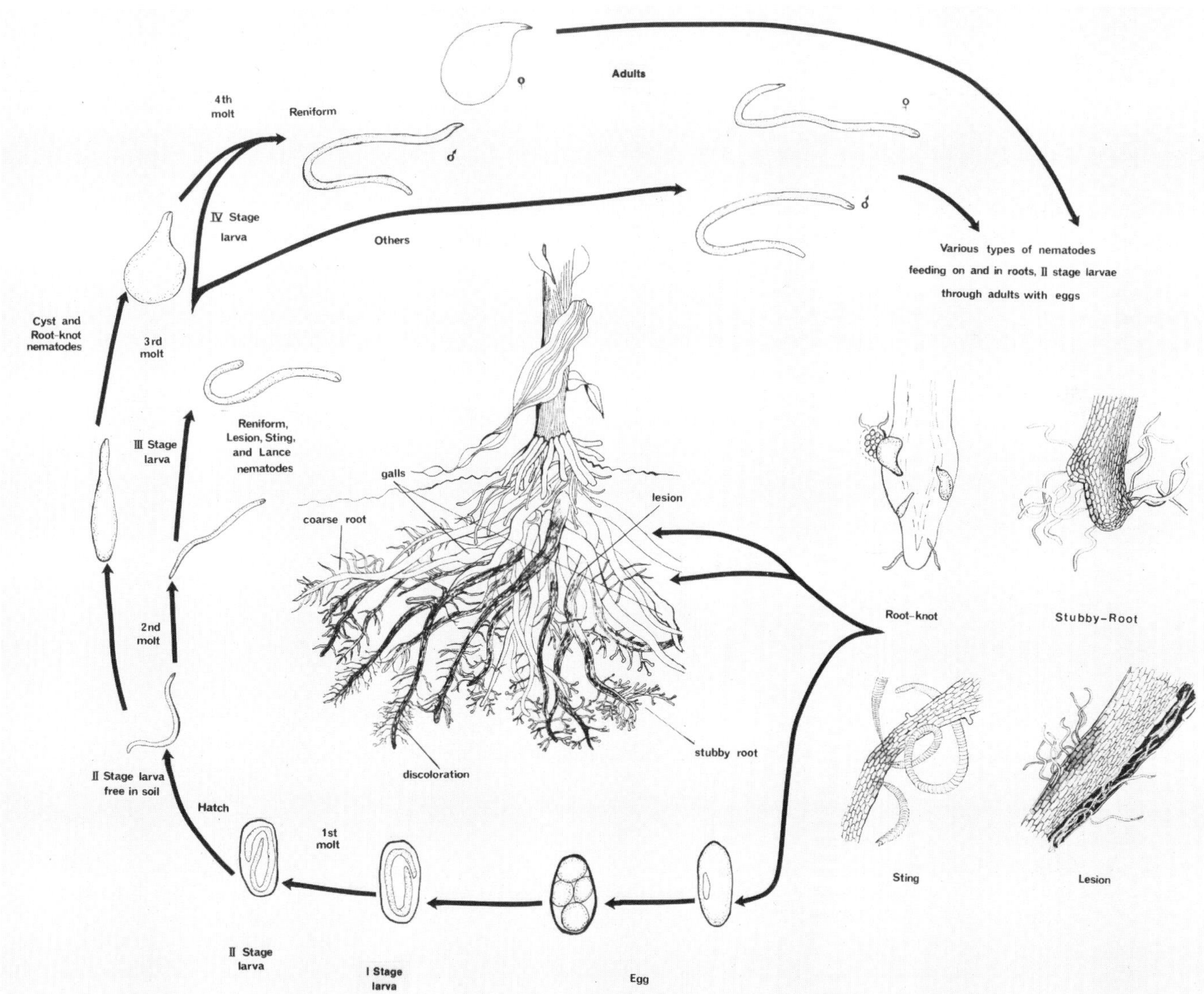

Fig. 74. Generalized life cycles of six genera of nematodes that feed on corn. (Drawn by Tina Ekstrom)

rarely be more than 30–75 cm per year.

Nematodes are extracted from soil by various flotation, sieving, and centrifugation procedures and from roots by incubation techniques. Normally, mixtures of parasitic and nonparasitic nematodes occur in, on, or about maize roots. Inferences about the pathological importance of each plant-parasitic species are drawn from knowledge of their feeding habits, relationship of above and below ground symptoms to numbers present, and frequency of occurrence. Inoculation of corn plants with individual species is usually necessary to demonstrate parasitism and the degree of pathogenicity.

Most plant-parasitic nematodes become inactive at temperatures below 10°C and over 35°C. The optimum temperature for activity of most nematodes is between 16 and 32°C, but varies greatly with the species and stage of development, growth of the host plant, and other factors.

More than 100 species of plant-parasitic nematodes have been reported to feed on or are associated with the roots of maize. Most corn plants in *every* field probably are attacked by one or more species. Feeding by these tiny parasites decreases the efficiency of the root system in absorbing and transporting water and nutrients to the top of the plant. Disruption of the normal root functions causes suppressed growth often accompanied by chlorosis or other nutrient deficiency, and ultimately lowered yields. The degree of damage is related to the cropping history of the field, the field preparation (tillage practices), the initial soil population of nematodes at planting time, and environmental conditions that develop after plant emergence. Usually maize plants are most susceptible to nematode injury when young and they become more tolerant with age. Soil hard pans, deficiencies, drought stress, and abnormally high temperatures weaken plants, rendering them more susceptible to damage from nematodes.

Virtually all fields will contain more than one species of plant-parasitic nematode. The kinds and numbers of nematodes present in the soil, relative to the age and condition of the plant and the environmental stress(es) on the maize plant, are important considerations in determining whether nematodes are causing economic injury to the crop. Damage often reflects the effects of the sum total of several species feeding on a root system and their interactions with other root-infecting pathogens.

The nematode genera *Longidorus, Xiphinema, Trichodorus,* and *Paratrichodorus* transmit viruses to various crop plants. Although these genera attack corn, this interaction has not been demonstrated with the known maize viruses.

Generalized Life Cycle

Life cycles of most plant-parasitic nematodes, although varying in details, are simple and direct (Fig. 74). Juveniles (larvae) that usually resemble the adults in structure and appearance hatch from eggs deposited by the female in the soil or within root tissue. Juveniles develop through four stages, each completed by a molt, and become adults after the final molt. Fully mature nematodes that reproduce either sexually or parthenogenetically lay eggs to start the next generation. Under optimum conditions, most plant-parasitic species complete their life cycles in three to four weeks, although some, such as the dagger nematode (*Xiphinema americanum*), may require nine months or longer to complete a cycle.

Feeding Habits

All nematodes that attack maize are obligate parasites that must feed on living plants to complete their life cycle. Almost all feed on plant cells by puncturing walls with a hollow stylet that resembles a minute hypodermic needle, injecting enzymes into the cells, and then ingesting the partially digested contents.

Plant-parasitic nematodes are divided into two major groups according to their parasitic behavior.

Ectoparasitic nematodes (e.g., *Xiphinema* and *Macroposthonia*) spend their life cycles *outside* the host and feed by inserting their stylets to varying depths within the root tissues. Endoparasites (e.g., *Meloidogyne* and *Pratylenchus*) feed both externally on and internally within plant roots, often becoming only partially imbedded in roots (semiendoparasites). Most forms (e.g., *Xiphinema* and *Pratylenchus*) remain mobile throughout their life cycles and "brouse" on roots. A few (e.g., *Meloidogyne*) become immobile assuming a sedentary parasitic habit (Figs. 74 and 81).

Nematodes damage the roots of maize in two major ways. Root cells are injured or killed mechanically or chemically (apparently through the secretion or injection of enzymes). Such damage renders the roots less efficient in absorbing and transporting water and nutrients. In many cases root growth ceases. Nematodes also interact with fungal or bacterial pathogens to cause disease complexes that often damage roots more severely than when either the nematode or other pathogen is present alone. Late in the season, nematode damage is difficult to separate from injury due to rootworms or other pests (Fig. 75).

Symptoms of Nematode Injury

Feeding of nematodes on maize roots reduces plant vigor and root growth, lowers the natural resistance of plants to other pathogens, and induces root lesions or galls and other deformations. Plant growth in heavily infested fields appears uneven, usually with distinct areas of stunted, chlorotic plants (Plate 55). The type

Fig. 75. Corn rootworm damage. **Left,** root system damaged by rootworms; **right,** healthy root system. (Courtesy R. R. Bergquist)

and severity of injury symptoms vary with the nematode species, population density, age and condition of the plants, and soil and climatic conditions. Commonly, damage caused by nematodes, especially widespread forms such as lesion nematodes, ranges from barely detectable to obvious reduction in yield. Losses may be localized in part of a field or spread over an entire field. Striking symptoms, however, often are *not* present.

In more severe cases, the primary symptoms on maize may include various combinations of the following:

1. Stunting: Suppressed plant growth. Plants are often uneven in height, usually in irregular spots, giving the field a "ragged" overall appearance (Plate 55).
2. Chlorosis or other discoloration: Yellowing similar to N, iron, magnesium, manganese, or zinc deficiency or purple coloration from P deficiency (Plate 56).
3. Wilting: Enhanced by wind and bright sunlight. Plants often slowly recover at night.
4. Root lesions: Dark, discolored areas in roots (Plate 57). With continued nematode feeding and invasion by other soil microorganisms, lesion size increases from pinpoint spots to large necrotic areas. Root-rotting and stalk-rotting fungi often invade tissues damaged by root-lesion nematodes, accelerating lesion enlargement (Fig. 76).
5. Root swellings: Indistinct swellings, primarily of the root tips, to knotlike galls throughout the root system (see Figs. 74, 79, 80, 81 and Plates 56 and 58) often accompanied by excessive production of branch roots ("hairy-root") above the swelling.
6. Injured or devitalized root tips: Cessation of root growth caused by feeding at or near the root tips. Tissues may retain normal coloration or turn brown and die. Devitalization frequently causes excessive production of roots near the soil surface.
   a. "Stubby roots"; root system composed of numerous short, stubby branches often arranged in clusters. (See Figs. 79, 80, 84 and Plate 56.)
   b. "Coarse roots"; root system with few or no branch roots or feeder rootlets. (See Figs. 80, 84, 85.)
7. Reduced and discolored root system: Overall reduction in the normal size of the root system accompanied by light to dark browning or blackening. (See Figs. 78, 79, 80, 81 and Plate 57.)
8. Yield reductions: Weakened plants produce smaller and fewer ears and kernels and less forage. Heavily damaged plants may produce no ears.

Diagnosis should *never* be made based solely on plant symptoms because symptoms of nematode damage often mimic those induced by other factors, such as low or unbalanced fertility, poor drainage, drought, soil insects, root-rot fungi, and herbicide damage (pages 79–

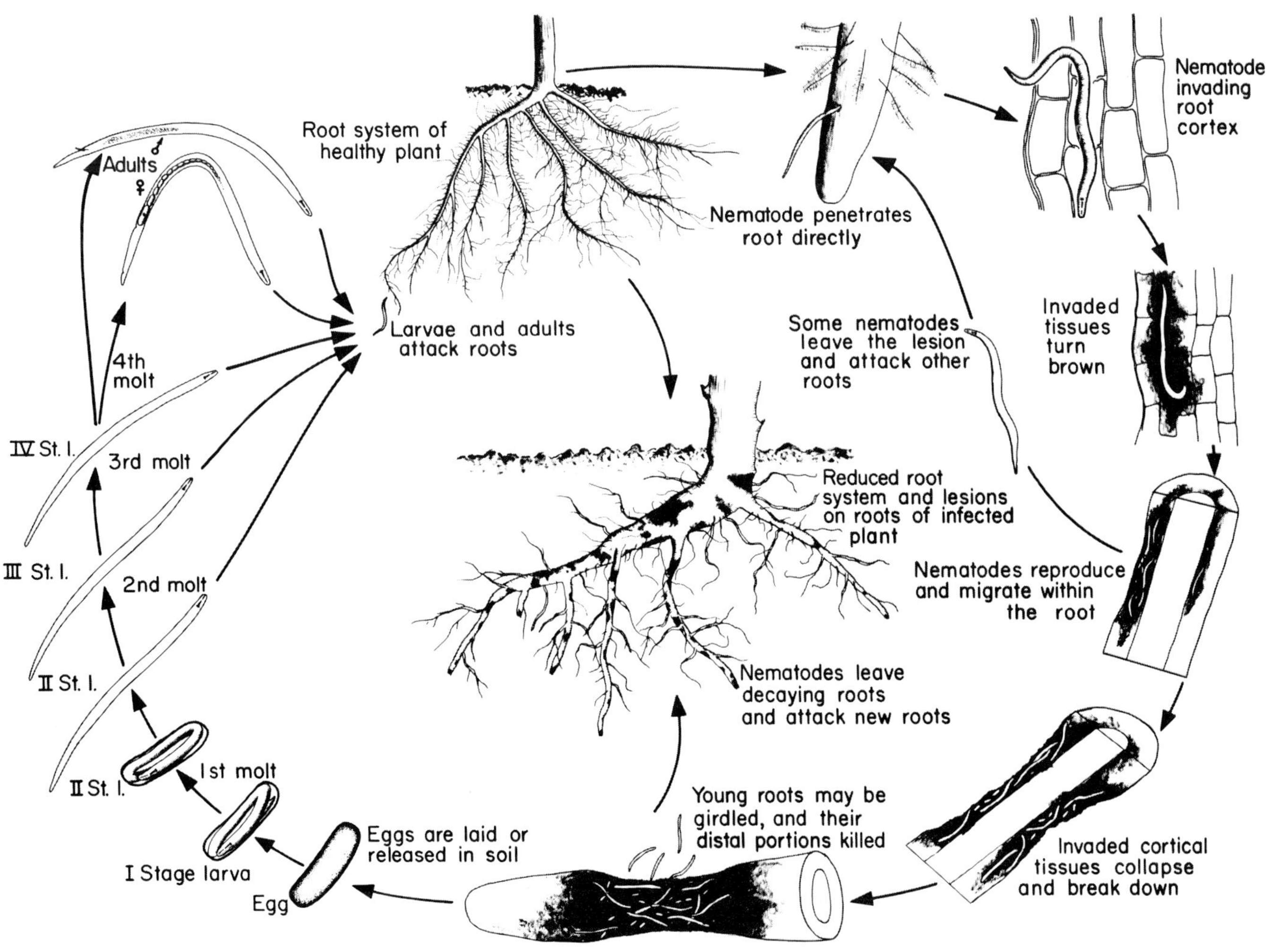

Fig. 76. Disease (life) cycle of root-lesion (*Pratylenchus* sp.) nematodes. (Reprinted, by special permission, from Plant Pathology, 2nd ed., by G. N. Agrios. ©1978 The Academic Press, New York)

87). Accurate diagnosis of nematode problems requires proper sample collection and handling of soil and roots and analysis by a competent nematologist.

Table VII summarizes most of the nematode diseases of maize.

Selected References

FERRIS, V. R., and J. M. FERRIS. 1974. Inter-relationships between nematode and plant communities in agricultural ecosystems. Agro-Ecosystems 1:275-299.

KRUSBERG, L. R. 1963. Host response to nematode infection. Annu. Rev. Phytopathol. 1:219-240.

MAI, W. F., and H. H. LYON. 1975. Pictorial Key to Genera of Plant-Parasitic Nematodes. 4th ed. (rev.). Comstock/Cornell Univ. Press: Ithaca, NY.

NORTON, D. C. 1978. Ecology of Plant-Parasitic Nematodes. Wiley-Interscience: New York.

SOUTHEY, J. F. (ed.). 1970. Laboratory Methods for Work with Plant and Soil Nematodes. Br. Ministry of Agric., Fisheries & Food. Tech. Bull. 2. HMSO: London.

THORNE, G. 1961. Principles of Nematology. McGraw-Hill: New York.

WALLACE, H. R. 1973. Nematode Ecology and Plant Disease. Crane, Russak & Co.: New York.

WEBSTER, J. M. (ed.). 1972. Economic Nematology. Academic Press: New York.

ZUCKERMAN, B. M., W. F. MAI, and R. A. ROHDE (eds.). 1971. Plant Parasitic Nematodes. Vols. I & II. Academic Press: New York.

Root-Lesion Nematodes

These nematodes (*Pratylenchus* spp.) are among the most common parasites of maize. Among the 12 species reported to attack the crop, *P. brachyurus, P. hexincisus, P. penetrans,* and *P. zeae* are well-known pathogens. Lesion nematodes are migratory endoparasites that feed and reproduce in the root cortex (Figs. 76 and 77). They are among the smallest plant-feeding nematodes, the adults ranging from 500 to 800 μm long. Some species, such as *P. brachyurus* and *P. zeae,* are more common and increase to large populations in sandy soils, primarily in warm regions. Others, such as *P. hexincisus* and *P. penetrans,* are common in heavy-textured soils and are most abundant in temperate regions. Corn is apparently tolerant to population densities of less than 500 per gram of dry root tissue

Table VII. Nematodes That Parasitize Maize Roots

Nematode	Mode of Parasitism	Characteristic Symptoms	Some Hosts other than Maize
Root-lesion (*Pratylenchus* spp.)	Endoparasitic	Poor root growth, necrotic root lesions, root decay, moderate stunting (Fig. 78 and Plate 57)	Grasses, cereals, sugarcane, legumes, tobacco, tomato, potato, strawberry, tree fruits, pines
Stubby-root (*Trichodorus* spp.; *Paratrichodorus* spp.)	Ectoparasitic	Stubby lateral roots, coarse roots, excessive upper roots, severe stunting, chlorosis (Plate 56)	Grasses, legumes, cotton, potato, tomato, cabbage, beet, citrus, ornamentals, many others
Sting (*Belonolaimus* spp.)	Ectoparasitic	Root lesions, stubby roots, coarse roots, severe stunting, chlorosis (Fig. 80)	Cereals, grasses, cotton, potato, cabbage, legumes, strawberry, celery, pines, many others
Root-knot (*Meloidogyne* spp.)	Endoparasitic	Root galls or excessive branch roots, severe stunting, chlorosis (Plate 58)	Legumes, cotton, tobacco, tomato, vegetables, strawberry, peach, ornamentals, many others
Lance (*Hoplolaimus* spp.)	Ecto-, semiendo-, and endoparasitic	Root lesions, moderate stunting, chlorosis (Fig. 83)	Grasses, legumes, sugarcane, cotton, pepper, tomato, citrus, pines, banana, many others
Dagger (*Xiphinema* spp.)	Ectoparasitic	Reduction of feeder roots, root decay, moderate stunting, chlorosis (Plate 55)	Grasses, legumes, strawberry, rose, tomato, trees, ornamentals, many others
Needle (*Longidorus* spp.)	Ectoparasitic	Stubby roots, thickened lateral roots, excessive upper roots, severe stunting, chlorosis (Figs. 84 and 85)	Grasses, potato, celery, grape, lettuce, many others
Spiral (*Helicotylenchus* spp.)	Ecto-, semiendo-, and endoparasitic	Reduction of feeder roots, root decay, mild stunting	Grasses, cereals, soybeans, peanut, clovers, tomato, strawberry, peach, pines, many others
Stunt (*Tylenchorhynchus* spp.)	Ectoparasitic	Poor root growth, moderate stunting, chlorosis (Fig. 86)	Grasses, cereals, tobacco, cotton, legumes, pepper, tomato, many others
Ring (*Macroposthonia* spp.; *Criconemoides* spp.)	Ectoparasitic	Root lesions, root decay, mild stunting	Grasses, citrus, apples, peaches, peanut, beans, soybeans, pines, many others
Awl (*Dolichodorus* spp.)	Ectoparasitic	Root lesions, stubby roots, coarse roots, stunting of plants	Legumes, tomato, pepper, celery, pines, pecan
Burrowing (*Radopholus similis*)	Endoparasitic	Necrotic root lesions, root decay, moderate stunting	Citrus, banana, avocado, sugarcane, rice, black pepper, tomato, hibiscus

when other conditions are favorable for plant growth, but populations of several thousand nematodes per gram are not uncommon. Under the latter conditions, plant damage is visible. Stunting, nutrient deficiency symptoms, chlorosis, and purple or reddish discoloration of the foliage may occur when dense populations are present. The reduced root system frequently is brown to black with few fibrous roots (Fig. 78). Host range reports of the important species of lesion nematodes include a wide diversity of agricultural crops and weeds.

Selected References

BERGESON, G. B. 1978. Control of the lesion nematode (*Pratylenchus* spp.) in corn with carbofuran. Plant Dis. Rep. 62:295-297.

EGUNJOBI, O. A. 1974. Nematodes and maize growth in Nigeria. I. Population dynamics of *Pratylenchus brachyurus* in and about the roots of maize and its effects on maize production at Ibadan. Nematologica 20:181-186.

ENDO, B. Y. 1959. Responses of root-lesion nematodes, *Pratylenchus brachyurus* and *P. zeae,* to various plants and soil types. Phytopathology 49:417-421.

Stubby-Root Nematodes

The relatively small, ectoparasitic, stubby-root nematodes (*Trichodorus* and *Paratrichodorus* spp.) are found mainly in sandy soils. They can increase rapidly to injurious levels and decline just as rapidly to barely detectable numbers in the soil. Feeding at or near root tips, these nematodes can be very damaging to maize by devitalizing the root tips and causing extensive "stubby-root" symptoms (Figs. 5, 79 and Plate 56). Severe plant stunting and chlorosis from N and magnesium deficiencies are common. Stubby-root and sting nematodes frequently occur together, the two forms comprising the most serious nematode pathogens of maize in the Southeastern USA.

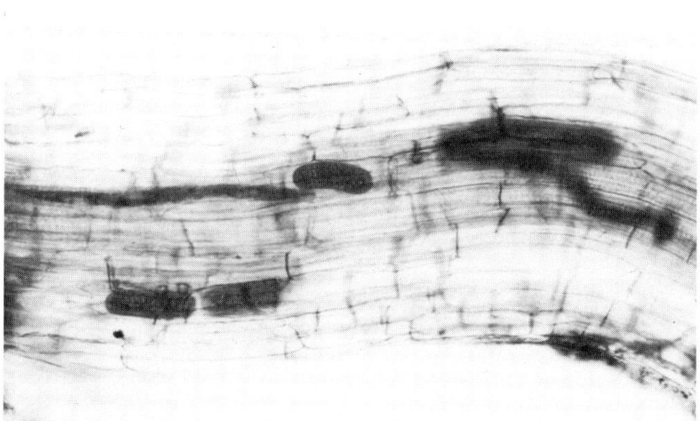

Fig. 77. Corn rootlet containing nematodes and eggs of *Pratylenchus hexincisus.* (Iowa Agricultural Experiment Station photo)

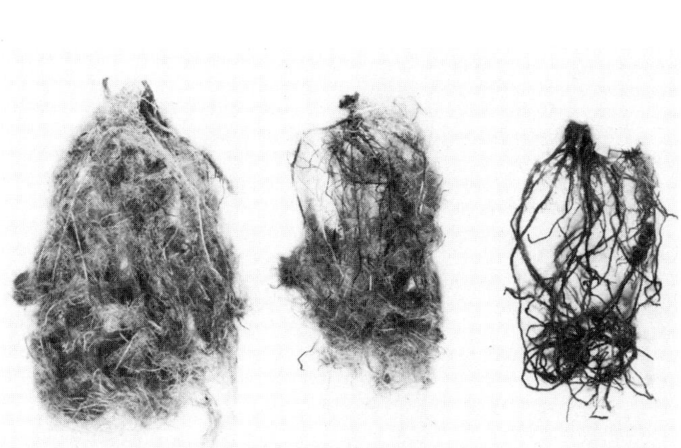

Fig. 78. Roots of corn growing in soil with **(left)** no *Pratylenchus hexincisus,* **(center)** soil moderately infested, **(right)** heavily infested with the nematode. (Iowa Agricultural Experiment Station photo)

Selected Reference

JOHNSON, A. W., C. C. DOWLER, and E. W. HAUSER. 1975. Crop rotation and herbicide effects on population densities of plant-parasitic nematodes. J. Nematol. 7:158-168.

Sting Nematodes

These large, brousing ectoparasitic nematodes (*Belonolaimus* spp.) are found only in sandy soils. They can be devastating, causing deep necrotic lesions in

Fig. 79. Stubby-root (*Trichodorus*) nematode injury. (Courtesy N. E. McGlohon)

Fig. 80. Sting nematode. Close-up of maize roots severely damaged by *Belonolaimus longicaudatus;* from greenhouse pot cultures artificially infested with hand-picked nematodes.

roots, devitalized root tips, coarse roots, and severe stunting (Fig. 80). Even small populations can cause serious damage, apparently because the nematodes produce a powerful phytotoxic enzyme while feeding. They are known to occur only in the USA, from Connecticut to northern Nebraska and Kansas and south to the Gulf and southeastern Atlantic Coast states, but are most serious in sandy coastal soils.

Selected Reference

RHOADES, H. L. 1978. Influence of nonfumigant nematicides and DBCP on *Belonalaimus longicaudatus* and yield of field corn in central Florida. Plant Dis. Rep. 62:91-94.

Root-Knot Nematodes

Several species of *Meloidogyne* attack maize and can cause severe damage, especially in warmer regions of the world. In tropical areas, they are usually the most serious of the nematode pathogens of maize. *M. incognita* seems to be the most important species, but occasionally *M. arenaria* and *M. javanica* can be serious. Root-knot nematodes are sedentary endoparasites that feed on specialized giant cells formed from vascular parenchyma tissue (Fig. 81). Females become swollen at maturity and are enclosed in galls on the roots. Eggs are deposited in a gelatinous sac on the surface of the gall. Root galls, hairy roots, and severe stunting are common symptoms (Plate 58). The host ranges of these nematodes are extremely wide and include over 2,000 kinds of plants. Corn germ plasm varies considerably in susceptibility. Thus, selection and use of resistant hybrids and varieties have good potential.

Selected References

BALDWIN, J. G., and K. R. BARKER. 1970. Host suitability of selected hybrids, varieties, and inbreds of corn to populations of *Meloidogyne* spp. J. Nematol. 2:345-350.
JOHNSON, A. W., and L. W. MORGAN. 1978. Influence of organic pesticides on nematodes, weeds, and insects and on yield of field corn. Ga. Agric. Exp. Stn. Res. Bull. 223.

Lance Nematodes

At least five species of the large, robust lance nematodes (*Hoplolaimus* spp.) have been associated with maize. *H. columbus* and *H. galeatus* are the most

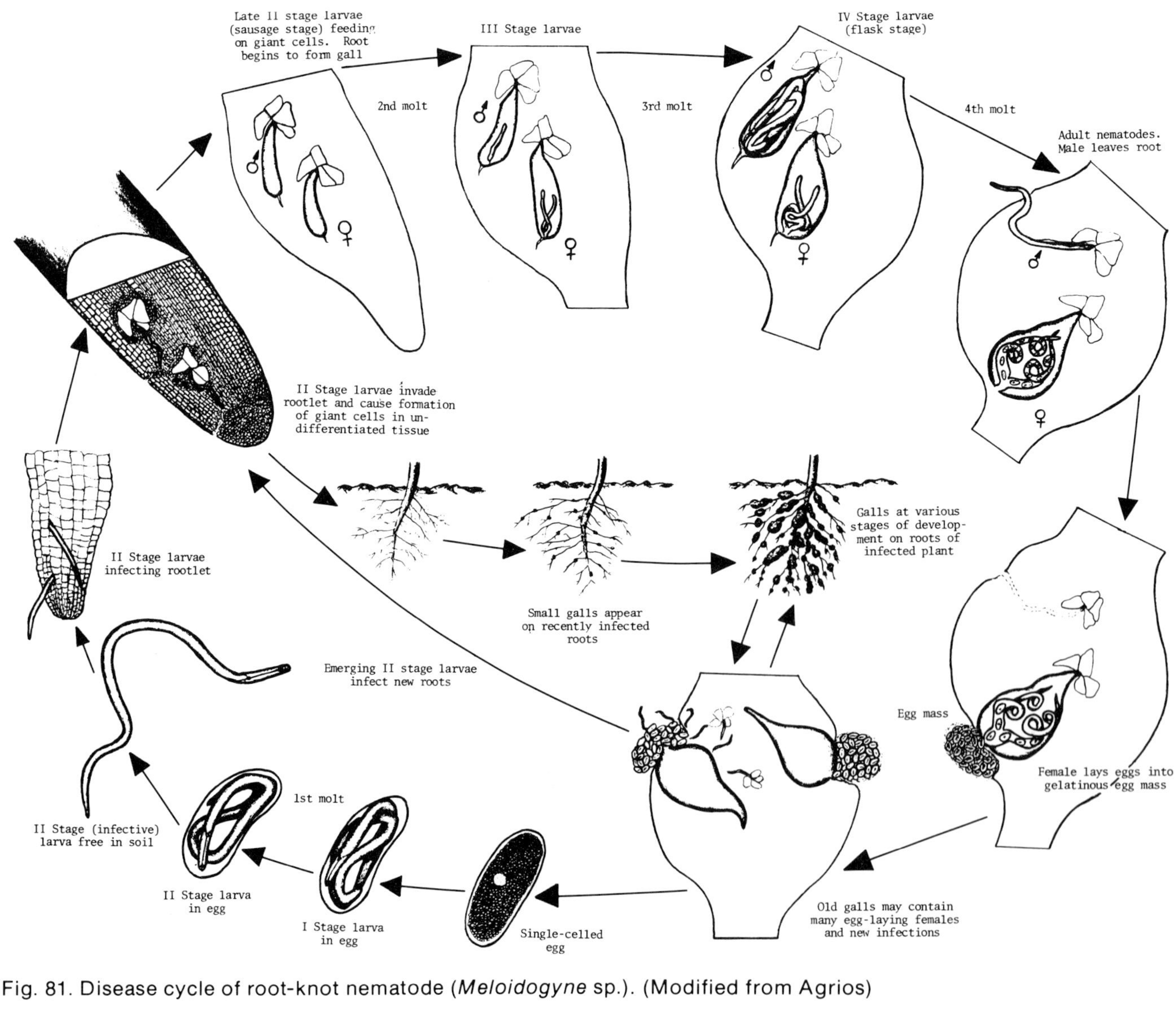

Fig. 81. Disease cycle of root-knot nematode (*Meloidogyne* sp.). (Modified from Agrios)

damaging species, causing root lesions that reduce root systems and stunt plant growth. *H. galeatus* is widespread on many crop and weed plants in North America and is particularly abundant on maize in sandy soils. This species is largely an endoparasite (Fig. 82) but sometimes feeds as an ectoparasite (Fig. 83) or partially imbedded in a root.

Selected References

LEWIS, S. A., and F. H. SMITH. 1976. Host plants, distribution, and ecological associations of *Hoplolaimus columbus*. J. Nematol. 8:264-270.

NORTON, D. C., and P. HINZ. 1976. Relationship of *Hoplolaimus galeatus* and *Pratylenchus hexincisus* to reduction of corn yields in sandy soils in Iowa. Plant Dis. Rep. 60:197-200.

Dagger Nematodes

In North America, *Xiphinema americanum* is one of the most common nematodes feeding on maize roots. Although pathogenicity has been demonstrated on only a few crops, such as alfalfa and red clover, this ectoparasite is generally considered to be an important pest, with an extremely wide host range. When populations are large, stunted plants (Plate 55) appear to be devoid of fine feeder roots. Dagger nematodes are common in both subtropical and temperate regions.

Selected Reference

FERRIS, V. R., and R. L. BERNARD. 1971. Effect of soil type on population densities of nematodes in soybean rotation fields. J. Nematol. 3:123-128.

Needle Nematodes

These nematodes (*Longidorus* spp.) can be among the most devastating pathogens of maize. They occur mainly in temperate regions around the world. *L. breviannulatus* is known to attack corn in Delaware, Illinois, and Iowa in the USA, but this species, along with others, is probably more widespread than reported. Members of this genus are among the longest plant-parasitic nematodes known, frequently attaining a length of 4–8 mm. Their occurrence, however, is often localized with dense populations occurring only in sandy soils where damage may be obvious soon after seedling emergence. Field symptoms include sparse stands of severely stunted, nutrient-deficient plants (Figs. 84, 85 and Plate 58). Root pruning is common, resulting in short, stubby, thickened, lateral feeder roots with slight swellings at the tips (Figs. 84 and 85).

Selected Reference

NORTON, D. C., and J. K. HOFFMAN. 1975. *Longidorus breviannulatus* n. sp. (Nematoda, Longidoridae) associated with stunted corn in Iowa. J. Nematol. 7:168-171.

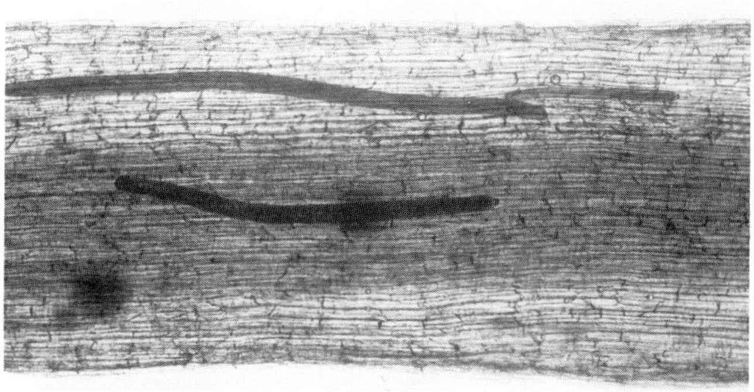

Fig. 82. Lance nematodes, *Hoplolaimus galeatus*, in a fibrous root of corn. (Iowa Agricultural Experiment Station photo)

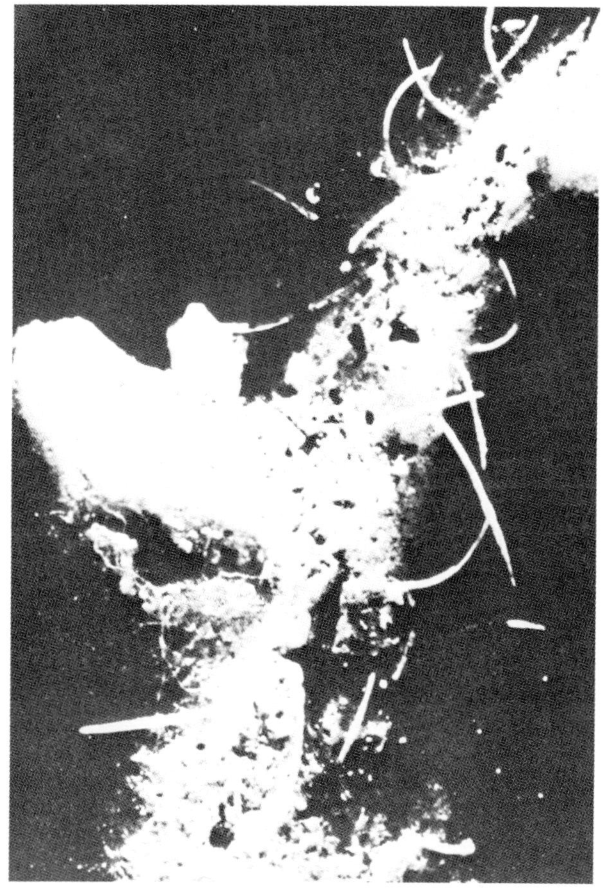

Fig. 83. Lance (*Hoplolaimus*) nematodes feeding on surface of maize roots. (Courtesy J. C. Wells)

Fig. 84. Corn roots heavily parasitized by the needle nematode, *Longidorus breviannulatus*. Note excessive branching and swellings at root tips. (Courtesy B. J. Jacobsen)

Fig. 85. Stunted corn plants moderately to severely damaged by the needle nematode, *Longidorus breviannulatus*. **Left,** stunting of plants. (Courtesy D. C. Norton) **Right,** coarse root system with little secondary root development. (Iowa Agricultural Experiment Station photo)

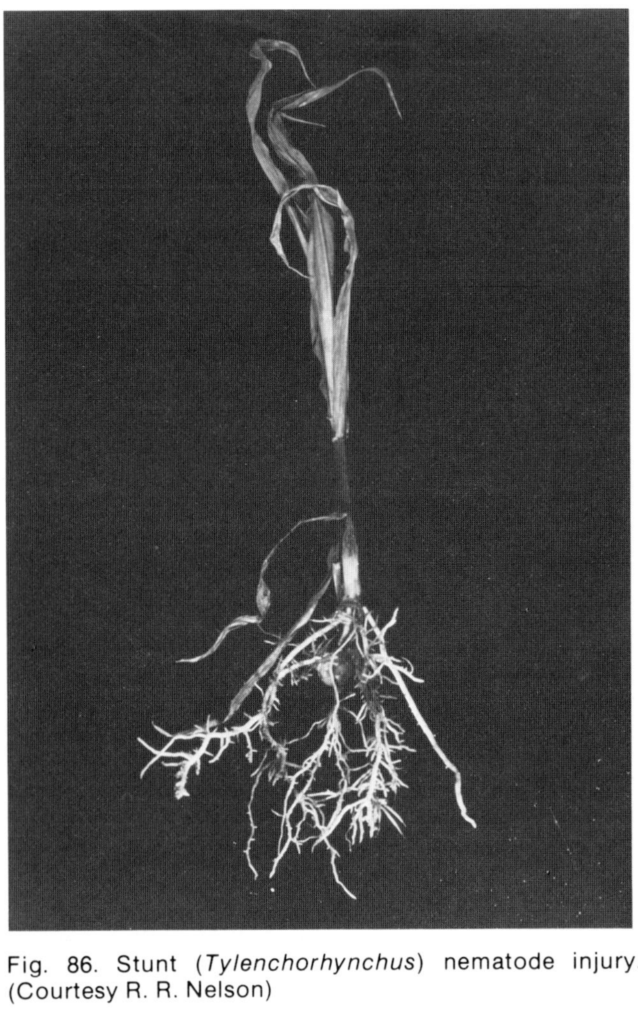

Fig. 86. Stunt (*Tylenchorhynchus*) nematode injury. (Courtesy R. R. Nelson)

Spiral and Stunt Nematodes

Numerous species of spiral and stunt nematodes (*Helicotylenchus, Tylenchorhynchus, Merlinius,* and *Quinisulcius* spp.) are commonly associated with maize (Fig. 86). Most fields contain one or more of these genera, which may be present in large numbers. Under controlled experimental conditions, however, unusually high numbers are required to demonstrate pathogenicity. Worldwide in distribution, these medium size nematodes are considered moderate pathogens at most, but they may contribute to overall damage. Stunt nematodes have highly brousing ectoparasitic feeding habits, while spiral nematodes may feed ectoparasitically or move partially or entirely into roots.

Selected References

NELSON, R. R. 1956. Resistance to the stunt nematode in corn. Plant Dis. Rep. 40:635-639.

NORTON, D. C. 1977. *Helicotylenchus pseudorobustus* as a pathogens on corn, and its densities on corn and soybean. Iowa State J. Res. 51:279-285.

Ring Nematodes

Macroposthonia and *Criconemoides* spp. represent one of the largest groups of plant-parasitic nematodes. Although the first species was described 100 years ago, their very feeding habit resulted in their being overlooked as important plant parasites until recently. Ring nematodes are small, stubby, sluggish ectoparasites which, though mobile, feed for long periods at one location practically anywhere on a root. Species such as

M. ornata have wide host ranges and cause damage on several crops. Although they frequently occur in abundance on maize, particularly in the Southeastern USA, the extent of damage by these nematodes to this crop has not been determined.

Selected Reference

JOHNSON, A. W., C. C. DOWLER, and E. W. HAUSER. 1974. Seasonal population dynamics of selected plant-parasitic nematodes on four monocultured crops. J. Nematol. 6:187-190.

Awl Nematodes

These generally ectoparasitic nematodes (*Dolichodorus* spp.) cause devitalized, coarse roots and stunting similar to sting nematode injury. Reports in the literature relative to maize are infrequent, and economic damage appears to occur primarily in the southeastern USA. They occur in limited areas, mainly in wet soils, but severe damage can result from relatively small populations.

Selected Reference

PERRY, V. G. 1953. The awl nematode, *Dolichodorus heterocephalus,* a devastating plant parasite. Proc. Helminthol. Soc. Wash. 20:21-27.

Burrowing Nematode

This nematode (*Radopholus similis*) is widely distributed in tropical and subtropical regions around the world, but is restricted in the USA to Florida. It can produce serious damage to corn, causing cortical lesions, root decay, and stunting of the plant, similar to injury from lesion nematodes. This endoparasite has a wide host range among crop and weed plants, but is most important on citrus and banana.

Miscellaneous Nematodes

Cyst nematodes (*Heterodera* and *Punctodera* spp.), cystoid nematodes (*Meloidodera* spp.), pin nematodes (*Paratylenchus* spp.), and reniform nematodes (*Rotylenchus* spp.) have occasionally been found parasitizing maize or are frequently associated with its roots. Their effects on growth and yield of maize, however, are unknown.

Control of Nematode Diseases

Resistant Hybrids and Varieties

Resistance or tolerance in maize is becoming available for only a few forms of nematodes. Progress in selection has been made mainly in the genera *Meloidogyne* and *Pratylenchus,* but the same hybrid or variety may respond differently to other species within each genus. Developing resistance to species in several nematode genera poses major problems.

Cultural Practices

Most nematodes that attack maize have extremely wide host ranges, thus complicating control by crop rotation. However, rotation with less susceptible or with nonhost plants, based on detailed nematode analysis, is often the most practical method of managing nematode populations to reduce damage to subsequent crops. Prevention of hardpan formation that restricts downward growth of roots often alleviates damage. Where feasible, alternate flooding and drying of fields at monthly intervals, or following with frequent plowing, can greatly reduce nematode populations. Much research is needed on the host ranges of the target nematodes and population dynamics before crop rotation and other cultural practices can be widely recommended.

Chemical Treatment

Soil fumigants are sometimes economically feasible with certain high-value crops in light, well-aerated soils, particularly where damage is severe and crop rotation cannot be practiced. Most fumigants must be applied preplant followed by a sufficient soil aeration period to avoid plant injury. Although usually not as effective as fumigants in reducing nematode populations, insecticide-nematicides, such as granular organophosphates and carbamates, are often practical in both sandy and heavy-textured soils. They are usually more economical than fumigants, and can be applied either preplant or at planting. Regardless of the type of nematicide, chemical treatment should not be undertaken without a preplant sampling and diagnosis to determine whether a potential nematode problem exists and to thus justify the expense of treatment. A limited trial treatment (strip tests) often is advisable to determine local response to nematicides where the problem is not well defined.

Parasitic Flowering Plants

Witchweed

This parasitic flowering plant is widely distributed, occurring in Africa, Australia, India, Southeast Asia, Indonesia, and in North and South Carolina (USA). Losses vary with the degree of field infestation, ranging from a trace to total.

Symptoms

Infected maize plants wilt and turn yellow producing symptoms resembling those caused by acute drought. Severe parasitism may cause stunting and death. One to several witchweed plants may grow above ground next to an infected maize plant. Below ground, fleshy stems of these and other witchweed plants that have not emerged are attached to the maize root system, and

Fig. 87. Witchweed (*Striga*) feeding on corn roots. (USDA photo)

infected roots bear witchweed haustoria. Roots of many witchweed plants may parasitize the same maize plant (Fig. 87).

Causal Organisms

Striga asiatica (L.) Kuntze (Syn. *S. lutea* Lour.) and *S. hermonthica* Benth.

S. asiatica is an obligately parasitic flowering plant with square, aerial stems and small, bright green, elongated leaves. The plant is erect, slender, rarely grows more than 15–30 cm tall, with opposite lower leaves and alternate upper leaves. Flowers are tubular, with a well-developed two-lipped corolla, 5–11 mm in diameter, and usually 10 calyx ribs. The flower color may be brick-red, yellowish red, lemon-yellow, or almost white with straw-yellow centers and yellow undersides (Plate 59). Seeds are minute (less than 0.25 mm long), striated, brown, and borne in pods or capsules. A single plant can produce up to half a million seeds.

S. hermonthica is generally larger than *S. asiatica* with bigger leaves and flowers with five calyx ribs. Its flowers are pink to pink-white.

Disease Cycle

Witchweed overseasons as seeds most of which require 90 days (for *S. asiatica*) to 15 or 18 months' dormancy before germination. The seeds germinate in the presence of a susceptible host or certain nonhost plants such as soybeans, peanuts, and cotton that produce a chemical stimulant necessary to trigger germination. When witchweed rootlets contact maize roots, the parasite establishes root connections by producing haustoria that penetrate the host roots within 8–24 hr by enzymatic or other chemical means. The parasite produces large numbers of minute seeds in pods or capsules and dies after seed production. From seed germination to seed production requires 60 to 90 days. The plants emerge about 30 days after germination and may produce flowers 14 to 20 days after emergence and the first mature seed pods approximately three weeks after flowering. However, viable seed is produced nine days after flowering on *S. asiatica*.

Epidemiology

The seeds are disseminated by contaminated soil on machinery, equipment, tools, water, or wind. They may lie dormant for 15 to 20 years. Growth of *S. asiatica* is inhibited in saturated, heavy soils. Plants are usually infested in a circular pattern with the center most seriously affected. The size of the infested area increases as the seeds are spread to surrounding areas. Witchweed has been reported parasitizing wheat, oats, barley, sorghums, sugarcane, rice, and more than 60 species in the grass and sedge families but is of economic importance only on maize, sorghums, sugarcane, Sudangrass, and rice. *S. hermonthica* occurs on sorghums and maize largely, if not entirely, in Africa.

Control

1. Prevent movement of *Striga* seed (in soil, plants, equipment) from infested to noninfested areas.
2. Keep all crops free of weedy grasses.
3. Apply appropriate herbicides to the base of maize stalks in hills or rows before *Striga* plants emerge or produce seed.
4. Plant nonhost trap crops (e.g., legumes) to stimulate germination of witchweed seeds. The parasite cannot invade legume roots and will starve to death if grassy weed hosts are not present. Usually a combination of methods is required for effective control.

Selected References

NELSON, R. R. 1957. Studies on the witchweed parasite of corn, pp. 19-24. Proc. 12th Annu. Hybrid Corn Industry Res. Conf.

NELSON, R. R. 1958. Preliminary studies on the host range of *Striga asiatica*. Plant Dis. Rep. 42:376-382.

ROGERS, W. E., and R. R. NELSON. 1962. Penetration and nutrition of *Striga asiatica*. Phytopathology 52:1064-1070.

SAUNDERS, A. R. 1933. Studies in phanerogamic parasitism with particular reference to *Striga lutea* Lour. S. Afr. Dep. Agric. Sci. Bull. 128.

Part II. Noninfectious or Abiotic Problems of Maize

Chemical Excesses and Injuries

Fertilizer Burn

Excessive amounts of soluble N or K fertilizer placed close to the seed may prevent germination. Seedlings often emerge and then die, or are severely stunted if they survive. Leaves turn yellow and sometimes die. Leaf tip necrosis is evidence of salt accumulation in the plant. Boron in the starter fertilizer is especially toxic to maize seedlings. *Favored by*: sandy soil, dry weather, fertilizer with a high salt index.

Anhydrous or Aqua Ammonia Damage

Short or long skips in rows scattered through the field may indicate anhydrous ammonia damage. In plants injured *after* emergence, roots appear to be sheared off or dried up at varying distances from the kernels. Leaves may be chlorotic and not fully expanded. The latter injury almost always occurs at the end of the field when ammonia (NH_3) is released while the knives are out of the soil (Plate 60). Fertilizers containing free NH_3 (e.g., diammonium phosphate), or which readily convert to NH_3 (urea), may be toxic if placed near the seed. *Favored by*: soil that is sandy, dry, compact, low in organic matter with an uneven surface; deep planting and shallow placement of NH_3.

Fig. 88. Onion leafing. (Courtesy W. O. Scott)

2,4-D Injury

Applications of 2,4-D may cause an "onion leafing" (Fig. 88) or a "buggy-whip" condition in which leaves remain wrapped in a tall spike. The herbicide may also cause malformation of the brace roots with some fasciation (Fig. 89) or growing together in a somewhat flattened condition. Brittleness of the stalks may result in breakage (see below). Injury may be due to excessive application rates, spraying in hot, humid weather, and spraying into the whorl. Broadcast application after the plants are over 30 cm tall can be injurious (Table VIII). Spraying near when the plants are silking or tasseling may affect kernel set and result in poorly filled ears.

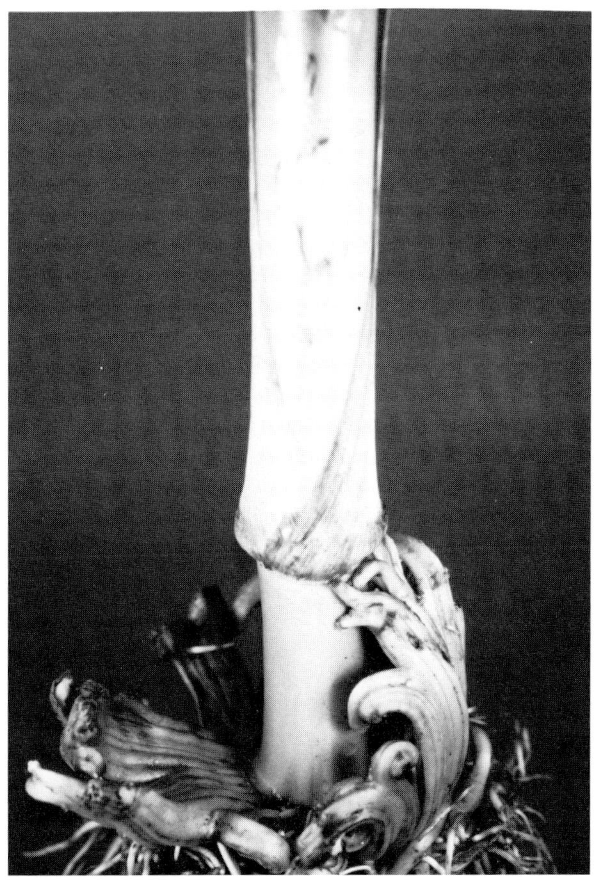

Fig. 89. 2,4-D injury to brace roots. (Courtesy J. P. Downs)

Injury is favored by the use of genetically susceptible hybrids.

Buggy-Whip

The characteristic symptom of this disorder is the slight or complete fusion of the upper part of the plant into a tight tube that encloses the upper leaves and tassel. The tube develops at the third to fifth node above the soil line, and is composed of thick and heavily lignified tissues that resemble a buggy-whip or an onion leaf (see 2,4-D injury above and Table VIII). Buggy-whip (Fig. 90) is often confused with genetic curling and infection by *Pseudomonas andropogonis* (bacterial stripe and leaf spot) or *Sclerophthora macrospora* (crazy top).

Fig. 90. Buggy-whip. (Courtesy D. C. Arny)

Table VIII. Herbicide Injuries to Maize

Herbicide (U.S. Trade Names)	Symptoms	Remarks	Control
Triazines Atrazine (several products) or atrazine plus oil or atrazine plus oil plus 2,4-D Cyanazine (Bladex), simazine (Princep), ametryn (Evik)	Gradual interveinal chlorosis. Leaves may die back from tips and turn light brown. Height of plants may be highly variable. Where severe, entire plants may be killed (Plate 61). Ametryn applied within 3 weeks of tasseling may affect pollination	Injury may occur on sandy soils low in organic matter or due to excessive rates, cool, wet weather, or other factors adversely affecting plant's metabolism	Use correct rate applied uniformly and accurately as label directs. Do *not* use cyanazine on sandy soils that are low in organic matter. Use at reduced rates in combination with another herbicide (see label)
Acetanilides Alachlor (Lasso), metolachlor (Dual), propachlor (Ramrod, Bexton)	Injured seedlings and older plants may have stunted or malformed shoots that fail to unfurl (Fig. 91). Plants may tend to leaf out underground; may have some "laddering"	Cool, wet weather prior to emergence is usually associated with injury, but majority of plants will outgrow damage	Use correct rate. Apply herbicide uniformly and accurately as label directs. Using dicamba (Banvel) with alachlor may accentuate problem. Hybrids differ in their susceptibility
Thiocarbamates Butylate (Sutan), EPTC (Eptam, Eradicane), vernolate (Vernam)	Extreme stunting, twisting, bending, and malformation of plants. Expanding leaves may rupture and shred. Ear malformation may occur (Fig. 92 and Plate 62)	A "safening agent" added to formulations reduces injury to maize. Hybrids differ in sensitivity	Use correct rate applied uniformly and accurately as label directs. Pay particular attention to method of incorporation
2,4-D (Several products)	*Preemergence applications* may produce severe stunting and malformation of roots and shoots. *Postemergence directed sprays* may cause fasciation and upcurling of the brace roots. Applied "over-the-top," especially on a hot day, 2,4-D may cause stalks to be brittle and break. Occasionally, a new leaf will fail to unfurl (onion-leaf) (Fig. 88). Applied near tasseling or at silking time, 2,4-D may interfere with seed set	Do not apply 2,4-D with atrazine and oil. Temperature and humidity at or near time of application, growth stage of maize, genetic susceptibility, method of application, and rates all influence amount of postemergence injury. To avoid breakage, avoid cultivating while plants are brittle	Use correct rate applied uniformly and accurately as label directs. If maize is over 20 cm tall, do *not* allow spray to enter whorl. Avoid spraying during cool, wet periods when maize is under stress or during hot, humid periods when corn is growing very rapidly. Do *not* spray between tassel and dough stages of growth. Avoid use of very susceptible inbreds, hybrids and varieties.
Benzoic acid Dicamba (Banvel)	Misapplication may cause onion leafing, proliferation of inhibited roots, abnormal brace root formation, or fasciation. Lodging may occur from postemergence applications. Resembles 2,4-D injury symptoms (see above)	Alachlor and dicamba combinations applied preemergence may cause injury, especially when unfavorable environmental conditions exist during seedling emergence	Use correct rate applied uniformly and accurately as label directs. Do *not* apply to sandy soils or soils low in organic matter. Create conditions favorable for rapid emergence and vigorous growth
Phenylureas Linuron (Lorox)	Injury is similar to that caused by triazine herbicides. Yellowing occurs first at leaf tips and margins followed by browning. Entire leaves may turn yellow and die (Plate 63)	This compound is taken up by the roots and translocated to the foliage. Can also affect leaves if applied postemergence	Use correct rate applied uniformly and accurately as label directs
Bipyridilium (Paraquat, gramoxone)	Leaf tissue turns brown and dies in flecks or spots where contacted by spray droplets; may superficially resemble bacterial or fungal leaf blight (Plate 64)	A nonselective contact herbicide with no significant soil activity. Not commonly recommended for postemergence use in corn	Use correct rate applied uniformly and accurately as a directed spray when maize is at least 25 cm tall. Expose no more than lower 7.5 cm of stalk to spray. *Control drift from adjacent fields*
Dinitroanilines Pendimethalin (Prowl), trifluralin (Treflan), dinitramine (Cobex), fluchloralin (Basalin), profluralin (Tolban)	Reduced stand, stunting, or uneven plant height, purpling of leaves with roots somewhat pruned and "clubby" at ends (Plate 65). Injury from pendimethalin may result from direct preplant application	Incorporation of pendimethalin increases risk of injury. May be confused with injury from nematodes (pages 69–77). Major damage is from carryover residue; chemicals differ in their persistence	Apply accurately and uniformly based on soil texture and organic matter level. Plant maize in soil treated with trifluralin, dinitramine, and other dinitroaniline herbicides *only* after sufficient time and rainfall has elapsed (see label)

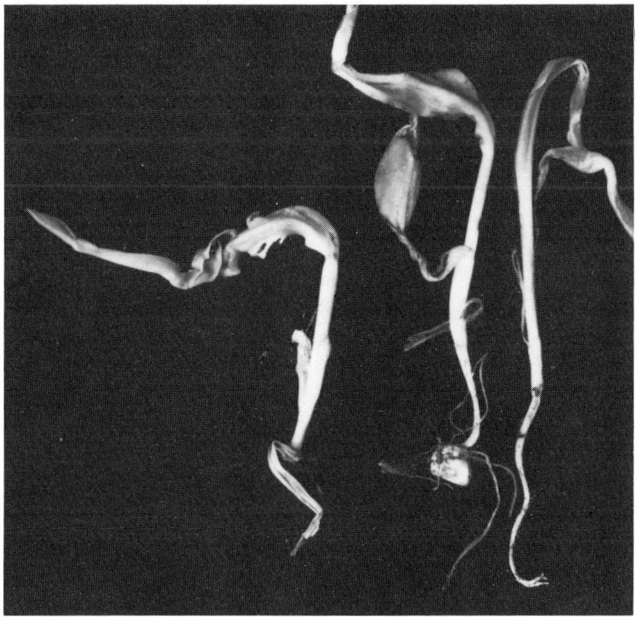

Fig. 91. Alachlor (Lasso) injury. (Courtesy E. L. Knake)

Fig. 92. EPTC (Eptam injury). (Courtesy E. L. Knake)

The disorder appears to be more prevalent in certain inbred lines grown in the central Corn Belt of the USA but is generally found only in small amounts in the field. When locally severe, yields can be reduced appreciably because affected plants are barren.

Buggy-whip may be caused by some herbicides such as 2,4-D. Buggy-whip predisposes affected plants to common smut (*Ustilago maydis*) infection.

Other Herbicide Injuries

Other preemergence or postemergence herbicides used at excessive rates, or chemicals to which maize has a low tolerance, may cause abnormal color, leaf spots, twisted or malformed shoots or roots, plants that never emerge, and stunting. The more common herbicide injuries, symptoms, and controls are summarized in Table VIII. Precision in application is the most important step in reducing chemical injury to a minimum. Application rates should be carefully selected and the equipment accurately calibrated. Maize hybrids and varieties differ considerably in their susceptibility to herbicide injuries.

Nutrient Excesses

In highly acid or alkaline soils, aluminum and iron oxides may accumulate and cause internal necrosis near shoot apices or in vascular bundles in the nodes. Nickel and chromium excesses may cause stunting, interveinal chlorosis, and puckering of the leaf margins.

Nutrient Deficiencies

Nutrient deficiencies are difficult to diagnose from visual symptoms alone since adverse weather, disease, insect, and various other injuries or other factors may cause similar manifestations. These symptoms are most reliable when all other factors favor rapid growth and generally occur on several plants in a broad area following a specific soil or management pattern (Table IX and center color spread). If possible, field diagnosis should be verified by plant analyses (chemical tissue tests) or soil tests. Oftentimes a careful appraisal of cropping history and fertility treatments will confirm the reasonableness of a tentative diagnosis. Nutrient deficiencies are least likely to occur where the crop is grown for a maximum yield within the pH range of 5.6 to not more than 7.3. Nutrient deficiency symptoms are most difficult to interpret in very cold or wet weather.

Nitrogen

N deficiency is characterized by stunted, spindly growth and pale, yellowish-green foliage in young maize plants. Older plants develop a V-shaped yellowing along the midrib to the tips of the bottom leaves. Later, the lower leaf tips die (firing) with the margins remaining green. Stalks remain thin and spindly. Ears tend to be pinched at the tips; kernels are glossy, hard, and flinty. *Favored by:* cold, wet, or flooded soils; drought (especially after midseason); sandy soils low in organic matter; heavy leaching rains; ponded areas when the weather is warm; and by intensive row-cropping with little or no forage legumes or animal manure.

Phosphorus

Mild deficiencies are characterized by stunted growth, whereas more severe cases cause purpling or reddening of the leaves beginning early in the growing season. Leaf tips die and turn dark brown. P deficiency is most pronounced in young plants. Ears on affected plants are small and often appear twisted with irregular kernel rows and with imperfectly developed ear tips. *Favored*

Table IX. Nutrient Deficiency Symptoms in Maize[a]

Color changes in *lower* leaves

Yellow, V-shaped discoloration beginning at midrib and widening toward leaf tip (*nitrogen*)
Purpling or browning from tip backward (*phosphorus*)
Brown discoloration and firing of leaf margins (*potassium*)
Yellow, interveinal discoloration, edges reddish purple (*magnesium*)
General chlorosis; tips of leaves stick together in ladderlike appearance (*calcium*)

Color changes in *upper* leaves

Young leaves show yellow to white bleached bands in lower parts (*zinc*) (Plate 66)
Interveinal chlorosis of entire leaf (*iron*)
Young leaves pale yellow, older leaves die at tips (*copper*)
General chlorosis of foliage (*sulfur*)
White, interveinal spots (*boron*)
Young leaves show pale green to yellow interveinal discoloration (*manganese*)
Young leaves wilt and die along margins (*molybdenum*)

[a]See "Guide to Nutrient Deficiency Symptoms."

by: cold, too wet or too dry soils; insufficient P either stored in or applied to the soil; restricted root growth in compacted soils; and root damage by insects, cultivator, fertilizer sidedressing knife, or herbicides.

Potassium

First symptoms are yellowing and dying of the leaf margins beginning at the tips of the lower leaves. Plants often lodge late in the season due to increased susceptibility to stalk rot. Ears on K-deficient plants may be small, chaffy, and dull with pointed, poorly developed tips. *Favored by*: sandy, organic, wet or compacted, strongly weathered soil and heavy K removal by the preceding crop or crops.

Calcium

Calcium is rarely deficient in maize. Leaf tips tend to stick together giving a ladderlike effect. Affected plants may have a slight yellowish-green tint and be severely stunted. *Favored by*: very low soil pH (below 5.5 in mineral soils and 4.8 in organic soils); high K and magnesium.

Magnesium

Deficiency manifests itself in seedlings as a general yellowing of the upper leaves. Eventually, yellow-to-white interveinal striping develops, and older leaves appear reddish purple along their edges and tips. Severe deficiencies cause leaf tips and margins to die, and the entire maize plant may show interveinal striping. Stunting may occur, but is not as severe as with P or K deficiency. *Favored by*: strongly acid, sandy soils in regions of moderate to high rainfall; high K; and soil treated with limestone low in magnesium.

Sulfur

Symptoms are stunting, delayed maturity, and a general yellowing of the foliage as in N deficiency. In some cases, interveinal chlorosis occurs, and yellowing is more pronounced on the younger leaves with sulfur deficiency than with N deficiency. *Favored by*: acid, sandy soils; low organic matter; and cold, wet soils.

Zinc

Symptoms occur usually within two weeks of seedling emergence, indicated by shortening of the internodes, lack of vigor, and light streaking of leaves followed by a broad stripe of bleached tissue on each side of the midrib (Plate 66). The midrib, margins, and leaf tip remain green. New leaves are sometimes almost white ("white bud" condition). Occasionally the leaf edges and interior of the stalks at the nodes appear purplish. Plants generally outgrow the deficiency unless it is severe. Very seldom is an entire field uniformly affected. *Favored by:* high P in soil or fertilizer; high soil reaction (above pH 6.3%); cool, wet soil; and low organic matter (exposed subsoil).

Iron

Upper leaves become pale green to almost white between the veins the length of the leaf. This deficiency is rare except on alkaline soils because maize has a low iron requirement. *Favored by*: high pH; wet, compacted, poorly aerated soils; and cool weather.

Manganese

Symptoms are vague because of the low manganese requirement of maize. Leaves become olive-green and may be somewhat streaked. If severe, deficient leaves have elongated whitish streaks the center of which turn brown, die, and fall out. *Favored by*: peat, muck, high pH, and sandy soils high in organic matter.

Copper

Copper (Cu) is rarely deficient in maize. The youngest leaves turn yellow as in iron deficiency. Severely deficient plants are stunted, younger leaves become chlorotic, and some dieback of older leaves occurs as in K deficiency. Stalks of affected plants are soft and flexible. Cu deficiency occurs mainly on organic soils. *Favored by*: very high organic matter content.

Boron

Boron is rarely deficient in maize. Symptoms are seen first on the youngest leaves as white, irregularly shaped spots between the veins. These spots may coalesce to form white stripes that appear waxy and raised from the leaf tissue. The internodes do not elongate and the tassels and ear shoots often fail to emerge giving plants a bushy appearance. *Favored by*: drought, high pH, and sandy soils low in organic matter.

Molybdenum

Molybdenum is rarely, if ever, deficient in maize. Older leaves die at the tip, along the margins, and finally between the veins. Younger leaves wilt, then die along the margins and in some cases twist. *Favored by*: low pH and strongly weathered soils.

Environmental Factors

Drought

Lack of moisture causes the upper leaves of young maize plants to roll upward and inward and appear dull (Plate 67). On young plants that have sufficient N but lack moisture, the leaves roll and become dull but do not turn yellow. In older plants, drought produces N deficiency even if enough N were applied. Severe stunting and irregular brown or tan patches of dead leaf tissue may develop, especially under marginal drought conditions. Seedlings may die, particularly in a loose, coarse seedbed and where the seed is located in moist soil, the first node in dry soil, and the mesocotyl rots (seedling blight). Dry surface soil, shallow planting, and lack of crown rot development result in poor emergence or young seedlings die.

Excess Moisture

The lower leaves of young plants in waterlogged soils turn yellow and die due to oxygen starvation when the roots are exposed to excess water. The crown area may die (Fig. 93). Plants up to 25 cm tall have a low tolerance to wet or flooded soils (Plate 68); plants with secondary roots stand "wet feet" much better.

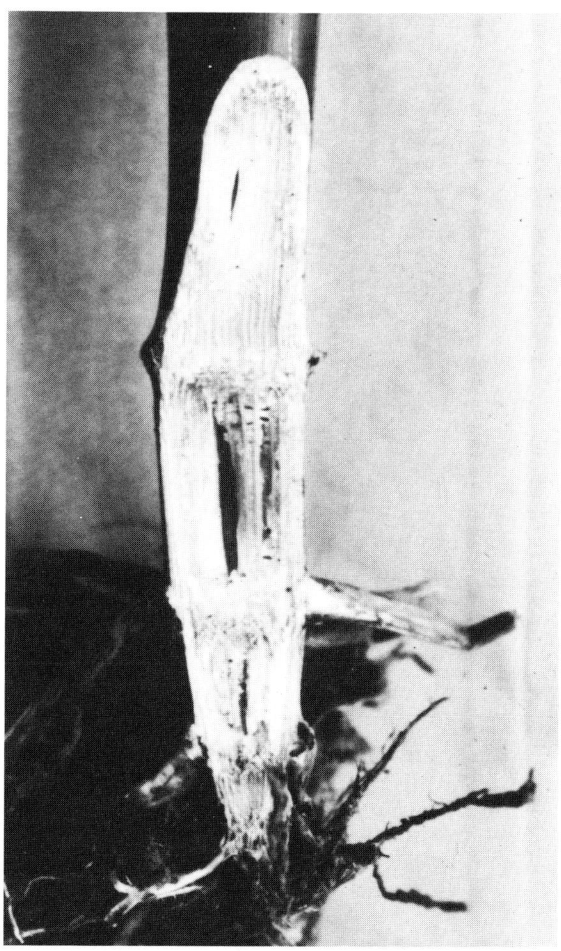

Fig. 93. Oxygen starvation of roots. (Courtesy R. R. Bergquist)

Soil Crust

Beating rains on fine-textured soil may produce such a hard crust that seedlings have difficulty penetrating it (Fig. 94).

Frost

Young maize plants are sometimes killed by a hard frost or freeze. Damage is most severe on organic (muck and peat) soils and is most common in low-lying areas. When the growing point is still underground, only the leaves are usually killed and plants normally recover

Fig. 94. Plant opening underground due to surface soil crusting. (Courtesy W. O. Scott)

Fig. 95. Poor seed set due to excessive heat during pollination. (Courtesy R. R. Bergquist)

Fig. 96. Low-temperature "sunscald" injury on a maize leaf. (Courtesy G. L. Worf)

(Plate 69). Bacteria (*Pseudomonas syringae* and *Erwinia herbicola*) serve as ice nuclei on the surfaces of seedling maize leaves. The severity of frost damage increases as the density of bacterial cells increases. Freezing of maize seeds reduces germination and vigor.

High Temperatures

High temperatures (38–52° C) cause scalding of maize leaf tissues, usually associated with low soil moisture and hot, dry winds. Terminal leaves are first chlorotic, then roll and bleach (Plate 70). Hybrids and varieties differ in their tolerance to high temperatures. During pollination, high temperatures may result in poor seed set or sterility (Fig. 95).

Low-Temperature "Sunscald" and Cold Banding

Injury occurs when chilly (5–15° C), clear, dewy nights are followed by clear, sunny mornings. Irregular, light gray to silvery blotches appear on both leaf surfaces of maize on the east side of affected plants (Fig. 96). The condition is quite common when the plants are 25–45 cm tall. Another symptom of low-temperature injury is the appearance of yellow bands across one or more leaves on young plants (Plate 71).

Wind

Strong winds can cause several types of damage to maize plants 10–25 cm tall. Leaf tips and edges turn white, then brown. Driving winds may carry sand and soil particles that cause abrasions on seedling leaves (Plate 72) or stalks on the windward side. Plants may also break or be injured below the soil line hampering growth. A broken ring may form around the hypocotyl just below the soil line. Such seedlings have purple or reddish-purple leaves as with P deficiency. Shredding of old and young plants occurs in areas where high winds prevail. Secondary roots are broken off (stubby) and do not become established in the soil. Plants lodged before complete establishment of the brace roots (and growth) may exhibit a bowed stalk that tends to make harvesting and walking between the rows difficult. Plants lodged after full attainment of growth may not recover.

Fig. 97. Hail injury. (Courtesy D. C. Arny)

Lightning

Lightning injury is rare on corn. When it does occur, plants in a small, round-to-oval area are killed. Damage is progressively less serious away from the point of discharge. The affected area has a sharp border.

Hail

Hail damage is easily recognizable by shredded leaves hanging in tatters (Fig. 97). Plants up to 25 cm may be completely broken or cut off. Nonetheless, such plants may produce a near-normal crop since the shoot apex is still below the soil line. Stalk rots or common smut may develop on the injured plants.

Other Noninfectious Factors

Purple Leaf Sheath

Irregular purplish-brown blotches and spots of varying size form on the leaf sheaths, usually after silking (Fig. 98). This harmless discoloration occurs when saprophytic fungi, such as yeasts and *Fusarium* spp., and bacteria develop on pollen and other particulate matter lodged between the stalk and leaf sheath, or following an infestation of aphids. The stalk under the sheath is *not* discolored or infected. The underside of the sheath is usually glossy, whereas it is velvet-like after an infection by the southern leaf blight fungus, *Helminthosporium maydis* Nisik. & Miy.

Genetic Stripe, Leaf Spots, and Flecks

Bright yellow to white stripes with *smooth* margins running the length of the leaf signify genetic stripe (Fig. 99). They may appear only on one side of a plant and usually on a few scattered plants in a field. Yellow-to-white spots and flecks may appear on corn leaves of all ages. These do not necessarily occur uniformly on a plant. Such patterns may suggest a disease, but the cause is genetic. (Not to be confused with bacterial leaf blight, Holcus spot, or Goss' bacterial wilt and blight, pages 5–11).

Barren Stalks and Red Leaf

Low fertility, drought, high plant populations, aphid or herbicide damage, chewed off silks, or poor timing of silking and pollen shed may all contribute to barren stalks or cobs without grain.

Red leaf may be caused by a broken midrib (Plate 73) or the accumulation of sugars and other photosynthetic products in leaves and sheaths. Similar accumulation often turns barren stalks red, although certain hybrids, commonly used as border rows by breeders, are naturally dark reddish to purplish.

Kernel Red Streak

Red streaking of the kernel pericarp (Fig. 100) is a symptom of kernel red streak (KRS). Streaked kernels are more common at the ear tips, particularly if the husks are loose and the kernels exposed. KRS is most common on yellow dent and least common on white maize, but may occur on sweet maize and popcorn. Streaks on the pericarp extend from the kernel base to the crown and vary from deep red in yellow kernels to pink-purple in white kernels. KRS is associated with toxin production from the feeding of a strain of the wheat curl mite, *Eriophyes* (*Aceria*) *tulipae* Keifer (Fig. 70), the vector of the wheat streak mosaic virus that affects maize (page 66).

Fig. 98. Purple leaf sheath. The irregular, purplish brown spots are of varying size and usually develop after silking.

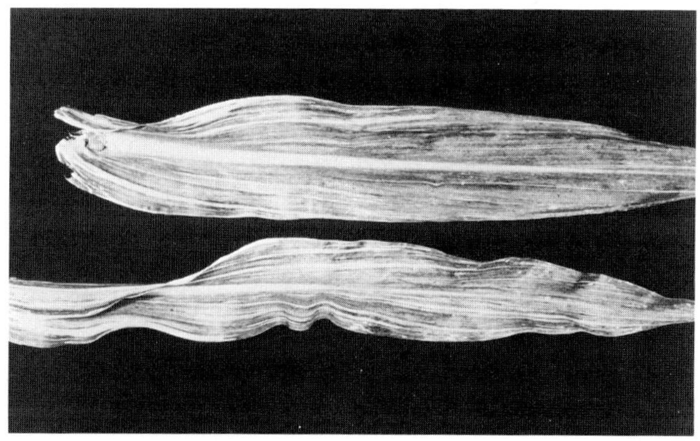

Fig. 99. Genetic stripe.

Fig. 100. Kernel red streak. (Courtesy G. L. Worf)

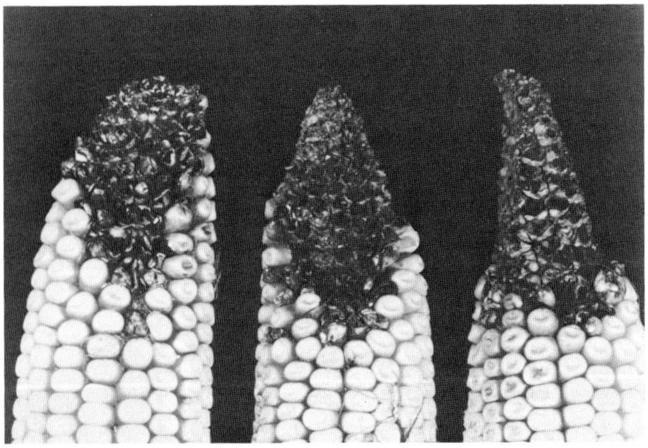

Fig. 101. Bird injury to the tips of maize ears.

Fig. 102. Popped kernel. (Courtesy A. J. Ullstrup)

Bird Injury

Bird injury to the tips of maize ears is shown in Fig. 101. This is more common in hybrids and varieties with erect ears and having exposed tips. Infection by numerous fungi commonly follows damage to the ears and grain.

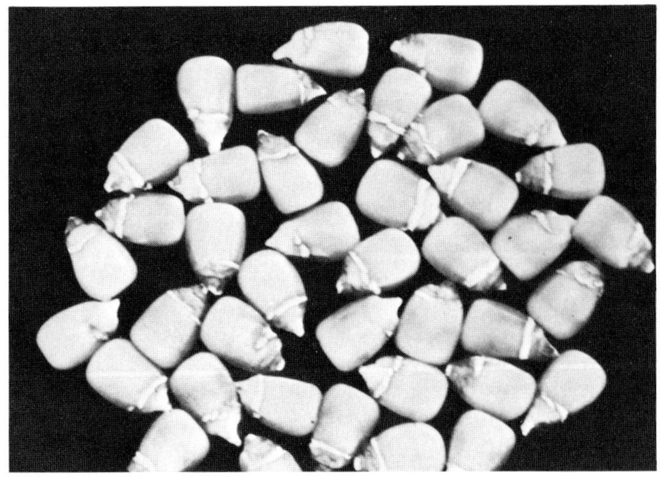

Fig. 103. Silk-cut. (Courtesy A. J. Ullstrup)

Exposed Ear Tips

This condition is most common where extreme drought occurs at kernel set, followed by abundant rainfall and good growing conditions. In extreme cases the ears may outgrow the husk cover.

Tassel Ears

A few scattered plants in a field may have combination tassels and ears in the same terminal structure. No specific cause is known for this condition.

Popped Kernel

Popped kernel appears as an irregular break in the seedcoat over the kernel crown. The kernels resemble partially expanded popcorn kernels (Fig. 102). This disorder is more common in a few inbred lines than in hybrids and varieties grown by farmers.

Silk-Cut

Despite its name, silk-cut has nothing to do with the silk of the ear. It appears as a horizontal cut or split in the pericarp over the sides of the kernel (Fig. 103). Silk-cut apparently affects only certain inbred lines and their combinations in hybrids and varieties.

Air Pollutants

Air pollutants cause substantial annual losses by visibly injuring leaves or suppressing plant growth without leaving visible injury. Pollutants may be either primary, emitted in a form toxic to plants (e.g., sulfur dioxide and hydrogen fluoride), or secondary (e.g., ozone and peroxyacetyl nitrate), resulting from photochemical reactions between nitrogen oxides and hydrocarbons in the air and energy supplied by sunlight.

In general, the extent and nature of air pollution injury depend on genetic and meteorological factors, as well as on the concentration and duration of the exposure to various pollutants singly or in combinations.

In examining plants for air pollution damage, consider the possible source of pollutants, terrain, previous weather patterns, abnormal symptoms on other vegetation in the area, time of symptom onset, as well as the possibility of mimicking symptoms due to living pathogens, insects, abnormal nutrition, or various environmental stresses.

Sweet maize is much more sensitive to air pollutants than field maize. Symptoms of fluoride toxicity and ozone injury are often similar to those produced by potassium deficiency or normal senescence. Injury by hormone-type herbicides (e.g., 2,4-D) is discussed in the section on chemical excesses and injuries (page 79).

Ozone

Symptoms on sweet maize are dull, gray-green, water-soaked lesions that develop into tan to white necrotic areas on both leaf surfaces. Small, silver-gray to tan, necrotic areas develop on some varieties in a band across one or both leaf surfaces. The oldest leaves may be injured at the base, the next older leaves across the middle, and the youngest leaves at the tip. Leaf margins are most severely injured, but the midrib remains undamaged (Plate 74).

Ozone injury may be confused with insect or mite damage, chlorine or fluoride injury, or natural senescence.

Fluorides

Sweet maize exposed to fluorides shows a chlorotic mottle or "flecking" along the leaf margins and tips. Small, irregular, chlorotic spots develop between the veins and may form continuous chlorotic bands. More severe injury causes interveinal and marginal chlorosis with some areas becoming necrotic (Plate 75).

Zinc or K deficiency, mite injury, genetic variation, or normal senescence cause symptoms resembling fluoride injury.

Chlorine

Chlorine injury to maize brings about interveinal, tan-yellow, necrotic streaks (Plate 76). Middle-aged or older leaves are usually more susceptible than young ones.

Sulfur Dioxide

Sulfur dioxide can injure maize plants of all ages. Acute injury appears as white necrotic areas between the veins. Acute or chronic injury results in tip dieback and premature senescence of leaves.

Peroxyacetyl Nitrate

Maize is generally considered resistant to peroxyacetyl nitrate (PAN) injury. Distinct bleached, chlorotic bands occur across the leaf blade. Young leaf tips may die back. If injury is severe, leaf cells in the band may be totally collapsed with the leaf folding downward at this point. Young, rapidly growing plants are more sensitive to PAN than older plants.

Effects of Environmental Factors

Conditions of light, temperature, humidity, soil moisture, and nutrition optimal for plant growth are also optimal for sensitivity to air pollutants. Most severe injury usually occurs during periods of air stagnation characterized as warm, clear, still, humid weather with high barometric pressure.

The following generalizations can be made:

1. Low temperatures for one to several days prior to exposure reduce sensitivity.
2. Plants growing under drought conditions are less sensitive to pollution injury.
3. Plants growing under relatively low fertility levels tend to have leaf injuries that mimic those caused by air pollutants.
4. The presence of plant pathogens on the leaves may affect sensitivity.
5. Sensitivity is affected by plant tissue maturity. Young leaves are usually free of injuries. Hybrids and varieties also differ markedly in their susceptibility to ozone and possibly other pollutants.

Selected References

ALDRICH, S. R., W. O. SCOTT, and E. R. LENG. 1975. Modern Corn Production (2nd ed.). A & L Publications: Champaign, IL.

ARNY, D. C., S. E. LINDOW, and C. D. UPPER. 1976. Frost sensitivity of *Zea mays* increased by application of *Pseudomonas syringae*. Nature 262:282-284.

ASHMAN, R. B., and A. J. ULLSTRUP. 1976. A heritable, noninfectious necrosis of maize leaves. J. Hered. 67:220-222.

HECK, W. W. 1968. Factors influencing expression of oxidant damage to plants. Annu. Rev. Phytopathol. 6:165-188.

HEGGESTAD, H. E. 1968. Diseases of crops and ornamental plants incited by air pollutants. Phytopathology 58:1089-1097.

HEGGESTAD, H. E., and W. W. HECK. 1971. Nature, extent, and variation of plant response to air pollutants. Adv. Agron. 23:111-145.

HOPPE, P. E. 1961. Sun scald of corn in Wisconsin in 1961. Plant Dis. Rep. 45:787.

INGLETT, G. E. (ed.) 1970. Corn: Culture, Processing, Products. Avi: Westport, CT.

JACOBSON, J. S., and A. C. HILL (eds.). 1970. Recognition of Air Pollution Injury to Vegetation: A Pictorial Atlas. Air Pollution Control Assoc.: Pittsburgh, PA.

JUGENHEIMER, R. W. 1976. Corn Improvement, Seed Production, and Uses. Wiley: New York.

KRANTZ, B. A., and S. W. MELSTED. 1964. Nutrient deficiencies in corn, sorghums, and small grains. In: Sprague, H. B. (ed.). Hunger Signs in Crops (3rd ed.). David McKay Co.: New York.

LINDOW, S. E., D. C. ARNY, and C. D. UPPER. 1978. *Erwinia herbicola:* A bacterial ice nucleus active in increasing frost injury to corn. Phytopathology 68:523-527.

NAULT, L. R., M. L. BRIONES, L. E. WILLIAMS, and B. D. BARRY. 1967. Relation of the wheat curl mite to kernel red streak of corn. Phytopathology 57:986-989.

NEUFFER, M. G., and O. H. CALVERT. 1975. Dominant disease lesion mimics in maize. J. Hered. 66:265-270.

NEUFFER, M. G., L. JONES, and M. S. ZUBER. 1968. The Mutants of Maize. Crop Sci. Soc. of America: Madison, WI.

Hosts for Maize Pathogens

Common Name	Scientific Name
African or finger millet	*Eleusine coracana* (L.) Gaertn.
Apple	*Malus sylvestris* Mill.
Avocado	*Persea americana* Mill.
Banana	*Musa X paradisiaca* L.
Barleys	*Hordeum* spp.
Barnyardgrass	*Echinochloa crusgalli* (L.) Beauv.
Beans	*Phaseolus* spp.
Beardgrasses or bluestems	*Andropogon* spp.
Beet	*Beta vulgaris* L.
Bentgrasses	*Agrostis* spp.
Bermudagrass	*Cynodon dactylon* (L.) Pers.
Black pepper	*Piper nigrum* L.
Bromegrasses	*Bromus* spp.
Broomcorn	*Sorghum bicolor* (L.) Moench (sens. lat.)
Broomcorn millet, corn millet, or wild proso millet	*Panicum miliaceum* L.
Bungalow grass	*Echinochloa stagnina* (Retz.) Beauv.
Cabbage	*Brassica oleracea* var. *capitata* L.
Celery	*Apium graveolens* L. var. *dulse* DC
Clovers	*Trifolium* spp.
Cordgrasses	*Spartina* spp.
Cottons	*Gossypium* spp.
Cowpea	*Vigna unguiculata* (L.) Walp. subsp. *unguiculata*
Crabgrasses	*Digitaria* spp.
Dallis grass	*Paspalum dilatatum* Poir
Eastern gamagrass	*Tripsacum dactyloides* L.
Eulalia	*Miscanthus japonica* Anderss.
Foxtails	*Setaria* spp.
Foxtail millet	*Setaria italica* (L.) Beauv.
Grain sorghum	*Sorghum bicolor* (L.) Moench
Grapes	*Vitus* spp.
Green foxtail or bristlegrass	*Setaria viridis* (L.) Beauv.
Guatemala grass	*Tripsacum fasciculatum* Trin. ex Aschers
Guineagrass	*Panicum maximum* Jacq.
Harding grass	*Phalaris stenoptera* Hack.
Hibiscus	*Hibiscus* spp.
Itchgrass	*Rottboellia exaltata* L.
Johnsongrass	*Sorghum halepense* L. Pers.
Lettuce	*Latuca sativa* L.
Lovegrasses	*Eragrostis* spp.
Lupines	*Lupinus* spp.
Merker grass	*Pennisetum setosum* (Swartz) L. Rich
Milo	*Sorghum bicolor* (L.) Moench (sens. lat.)
Molasses grass	*Melinis minutiflora* (Beauv.)
Oats	*Avena* spp.
Orchardgrass	*Dactylis glomerata* L.
Paragrass	*Brachiaria mutica* (Forsk.) Stapf.
Peach	*Prunus persica* L. Batsch.
Peanut	*Arachis hypogaea* L.
Pearl millet or bajra	*Pennisetum glaucum* L. R. Br.
Pepper	*Capsicum frutescens* L. or *C. annuum* L.
Pines	*Pinus* spp.
Pitscalegrass	*Hackelochloa granularis* (L.) Kuntze
Plumegrasses	*Erianthus* spp.
Potato	*Solanum tuberosum* L.
Quackgrass	*Agropyron repens* (L.) Beauv.
Red fescue	*Festuca rubra* L.
Redtop	*Agrostis alba* L.
Rice	*Oryza sativa* L.
Roses	*Rosa* spp.
Rye	*Secale cereale* L.
St. Augustine grass	*Stenatophrum secundatum* (Walt.) Kuntze
	Setaria vulpiseta (Lam.) R. & S.
Shattercane	*Sorghum bicolor* (L.) Moench (sens. lat.)
Sorghum	*Sorghum bicolor* (L.) Moench (sens. lat.)
	Sorghum arundinaceum (Willd.) Stapf.
	Sorghum propinquum (Kunth) Hitchc.
Soybeans	*Glycine max* (L.) Merr.
Strawberries	*Fragaria* spp.
Sudangrass	*Sorghum sudanense* (Piper) Stapf.
Sugarcane	*Saccharum officinarum* L.
Sugarcane, wild	*Saccharum spontaneum* L.
Tanglehead	*Heteropogon contortus* (L.) Beauv.
Teosinte, annual	*Zea mays* L. subsp. *mexicana* (Schrad.) Iltis
Teosinte, perennial	*Zea perennis* (Hitchc.) Reeves & Mangelsd.
Timothy, common	*Phleum pratense* L.
Tobaccos	*Nicotiana* spp.
Tomato	*Lycopersicon esculentum* Mill.
Vasey grass	*Paspalum urvillei* Steud.
Wheats	*Triticum* spp.
Witchgrass	*Panicum capillare* L.
Woodsorrels	*Oxalis* spp.

Glossary

cc—cubic centimeter
cm—centimeter (0.39 inch; 10 millimeters = 1 cm)
g—gram (453.59 grams = 1 pound)
ha—hectare (1 hectare = 2.47 acres)
hr—hour
in.—inch (2.54 centimeters)
kg—kilogram (1 kilogram = 2.2046 pounds)
m—meter (1 meter = 39.37 inches)
μ—micron
μg—microgram; 10^{-6} grams
μm—micrometer; 10^{-6} meters (1.001) or 0.00003937 of an inch
mm—millimeter (10 millimeters = 1 centimeter; 0.4 inch)
nm—nanometer; 10^{-9} meters

acervulus (pl. **acervuli**)—erumpent, saucer-shaped, cushion-like fruiting body of a fungus bearing conidiophores, conidia, and sometimes setae
acid-fast—bacterial cells that are not decolorized by 25% sulfuric acid after staining with carbol fuchsin
acid soil—soil with an acid reaction (pH less than 7)
actinomycetes—filamentous bacteria, "ray fungi"
acute—as opposed to chronic; symptoms that develop suddenly
adventitious—arising or occurring sporadically or in other than the usual place
aeciospore—dehiscent, binucleate, dikaryotic spore produced in an aecium of a rust
aecium (pl. **aecia**)—more or less cup-shaped sorus in the rusts; a "cluster-cup" that produces chains of aeciospores
aflatoxin—series of toxic metabolic products of *Aspergillus flavus* and related fungi; harmful to man and animals; the cause of a fatal disorder (aflatoxicosis) in certain animals, e.g., young turkeys, ducklings, and rainbow trout; carcinogenic for rats and possibly man.
alkaline soil—soil with a basic reaction (pH greater than 7)
alternate host—one of two species of plants on which a parasitic fungus (e.g., a heteroecious rust) must develop to complete its life cycle
anastomosis (pl. **anastomoses**)—fusion between branches of the same or different hyphae (or other structures) to make a network
antheridium (pl. **antheridia**)—male sex organ (male gametangium) in fungi
anthracnose—leaf or stem disease caused by fungi (Melanconiales) that form acervuli; characterized by sunken lesions and necrosis
antibiotic—damaging to life; a chemical usually of microbial origin that inhibits or kills other microorganisms
antigen—normally a protein substance, but any foreign chemical that induces antibody formation when introduced into living animal tissues
apex (adj. **apical**)—tip
apothecium (pl. **apothecia**)—cuplike or saucerlike ascus-bearing fruiting body (ascocarp) in the Discomycete section of the Ascomycetes
appressorium (pl. **appressoria**)—swelling on a germ tube or hypha, especially for attachment to a host in an early stage of infection, as in certain rusts and *Colletotrichum*
ascocarp—sexual fruiting body (ascus-bearing organ) of an Ascomycete fungus, i.e., apothecium, cleistothecium, perithecium
ascogenous—developing or originating from an ascus
ascomycete—fungus that produces sexual spores (ascospores) within an ascus
ascospore—spore borne in an ascus by "free cell formation"
ascus (pl. **asci**)—saclike or clavate cell containing ascospores (typically eight) and borne in an ascocarp
aseptate—without cross-walls
asexual—vegetative; without sex organs, gametes, or sexual spores; imperfect state of a fungus
autoecious—rust fungus that completes its life cycle on one host
avirulent—without capacity to cause disease; nonpathogenic

bacilliform—like a blunt, short, thick rod
bactericide—chemical or physical agent that kills or protects from bacteria
bacteriocin—antibiotic-like substance produced by some bacteria active against strains of the same or related bacterial species
basidiocarp—sexual fruiting body (basidium-bearing organ) of a Basidiomycete fungus
basidiomycete—fungus characterized by septate mycelium, sometimes with clamp connections, that forms sexual spores (basidiospores or sporidia) on a basidium
basidiospore—haploid (n) spore produced on a basidium
basidium (pl. **basidia**)—short, club-shaped, haploid (n) promycelium produced by Basidiomycete fungi (smut and rust fungi included) on which basidiospores are borne
Bi-(prefix)—twice, having two; **-ciliate**, having two cilia; **-ennial,** plant having a life span of more than one year but not more than two years; **-flagellate,** having two flagella; **-guttulate,** with two globules or vacuoles; **-nucleate,** having two nuclei; **-polar**, at the ends; **-seriate**, in two series; **-tubular,** having a tube within a tube; **-tunicate** (of asci) having two walls; **-verticillate,** having branching at two levels, as in some species of *Penicillium*
biological control—disease or pest control through counterbalance of microorganisms and other natural components of the environment
biotype—physiologic race; subspecies group of organisms that differ in biochemical, physiological, or behavioral properties; a group of individuals having a like genetic makeup. *See physiologic race*
blade—flat or expanded part of a leaf or petal
blight—general term for sudden, severe, and extensive spotting, discoloration, wilting, or destruction of leaves, flowers, stems, or the entire plant. Usually young growing tissues are attacked. May be coupled with the name of the host plant attacked, e.g., leaf blight, tip blight
blotch—usually a superficial irregularly shaped spot or blot
bract—modified leaf associated with a flower or inflorescence
broadleaf—any plant with a flat leaf
bursa—lateral cuticular extension at the posterior of certain male nematodes

C—centigrade or Celsius
calyx—outermost flower whorl; sepals collectively
capsid—protein coat of a virion
carbohydrate—food compound composed of carbon, hydrogen, and oxygen, with hydrogen and oxygen in the ratio of 2 to 1
carcinogen—a substance or agent producing or inciting

cancer

carrier—plant or animal harboring a parasite (e.g., virus) without itself showing symptoms of disease. Carrier plant can be a source of infection to other plants. An insect or mite contaminated externally with an infectious agent (e.g., bacterium, virus, fungus, nematode) is sometimes called a carrier

catalase—enzyme of qualitative significance in some bacteria

causal agent—organism or agent that produces a given disease

chitin—carbohydrate in fungal cell walls

chlamydospore—thick-walled or double-walled, asexual resting spore formed (terminal or intercalary) from hyphal cells or by transformation of conidial cells; also used for smut spores (more correctly, teliospores)

chlorophyll—green, light-sensitive pigments, found chiefly in chloroplasts of higher plants, that participate in photosynthesis

chlorosis (adj. **chlorotic**)—failure of chlorophyll development due to nutritional disturbance (e.g., lack of iron, zinc, or magnesium) or disease; fading of green plant color to light green, yellow, or white

chronic—slow-developing, persistent, or recurring symptoms

chytrid—microscopic, usually aquatic, nonmycelial fungus member of the Chytridiales

cilium (pl. **cilia**)—flagellumlike appendage, especially of zoospores, which propels them through water; hairlike projection from certain cells, e.g., from the edge of an apothecium

circulative—property of viruses that accumulate within or pass through the gut of insect vectors before transmission to plants

cirrus, cirrhus (pl. **cirri**)—ribbonlike cylinder of spores held together by mucus as it issues from an ostiole; a tendril-like mass or "spore horn" of forced-out spores

clavate—club-shaped, narrowing in the direction of the base

cleistothecium (pl. **cleistothecia**)—closed, usually spherical ascocarp; typical of the powdery mildews (Erysiphaceae)

coalesce—to join together, merge, overlap

coenocytic—nonseptate; refers to nuclei being embedded in the cytoplasm without being separated by cross-walls

coleoptile—ephemeral, nonpigmented tissue sheathing the first true leaf of a grass (or maize) seedling

collateral host—secondary or subordinate host

complete fertilizer—any fertilizer containing the three basic elements usually lacking in soil: nitrogen (N), phosphorus (P), and potassium (K)

confluent—coming together

conidiophore—simple or branched hypha (a fertile hypha) on which conidia are produced

conidium, conidiospore (pl. **conidia**)—any asexual spore borne at the tip or side of a specialized hypha, called a conidiophore; any asexual spore except intercalary chlamydospores or sporangiospores

corolla—collectively the petals of a flower

cortex (adj. **cortical**)—tissues between the epidermis and phloem in stems and roots; a more or less thick outer covering; peridium

cover crop—plants grown to improve and maintain soil structure, add organic matter, and prevent soil erosion

crazy top—disease symptom manifested by twisting, proliferation, and distortion of upper plant parts caused by a downy mildew fungus

cross protecting—property of a pathogen (usually a virus) established in a host that prevents or inhibits the establishment of a second pathogen in the same host

crown—compacted series of nodes from which shoots and roots arise

culm—stem of a grass plant

cultivar—cultivated variety

culturing—artificial growth and propagation of organisms on nutrient media or living plants

cuticle—waxy noncellular layer formed on outer walls of plant epidermal cells; outer sheath or membrane of a nematode or plant

cyst—egg-laden carcass of a dead female nematode of the genus *Heterodera*

cyt- or **cyto-** (prefix)—of or pertaining to a cell; **-ology**, science of cell structure and function; **-plasm**, all the living protoplasm in a cell, except the nucleus

damped-off, damping-off—destroyed near soil level; said of germinating seed or seedlings that suffer a rapid, lethal decline before or after emergence

dehiscent, dehiscing—opening when mature by pores or by breaking into parts; splitting along definite seams at maturity

desiccate—to dry

diagnosis—identification of the nature and cause of plant trouble

dichotomous, dichotomy—dividing into two parts; type of branching in plants in which the main axis forks repeatedly into two more or less equal arms

dieback—progressive death of shoots, leaves, or roots beginning at the tips

digitate—having fingerlike divisions

dikaryon (adj. **dikaryotic**)—fungal cells having two sexually compatible nuclei per cell (n+n)

diplo- (prefix)—two, twice, double; **-karyon** or synkaryon; **-phase**, the part of a life history in which the cells are diploid

diploid—(of a nucleus), having the 2n number of chromosomes; (of a cell), having the 2n number of chromosomes in one (synkaryotic 2n) or two (dikaryotic n+n) nuclei; (of basidiomycete mycelium), made up of dikaryotic diploid cells

dissemination—spread of infectious material (inoculum) from a diseased to a healthy plant by wind, water, man, insects, mites, machinery, etc.

distal—far from the point of attachment or origin

diurnal—daily; during the day as opposed to nocturnal

DNA—deoxyribonucleic acid

dormant—resting; living in a state of reduced physiological activity

dorsal—back or upper surface as opposed to ventral

dwarfing—symptom of disease; underdevelopment of a plant or plant organs. May be caused by any type of pathogen, faulty nutrition, unfavorable environmental conditions, etc.

ear—fruiting spike of a cereal (as maize) including the kernels and protective structures (husks)

echinulate—(of spores, etc.) having small, pointed spines projecting from cell walls

ecto- (prefix)—outside; **-parasite**, a parasite living on the outside of its host; **-spore**, an exogenous spore; **-trophic** (of mycorrhiza), covering the root surface

edema (adj. **edematous**)—a watery swelling of plant organs or parts; an abnormal accumulation of serous fluid in connective tissue or in a serous cavity

eelworm—nematode

effuse—stretched out; flat and spreading

embargo—a government order prohibiting transport of plants or plant parts into a country or region to prevent spread of disease, pathogens, or vectors

embryo—seed "germ"; the rudimentary plant within a seed

emetic—agent that induces vomiting

enation—small swelling or gall

encyst—form a cyst or encasement

endemic—natural to one country or geographical region

endo- (prefix)—inside; **-biotic,** growing inside living organisms; **-conidium**, a conidium formed inside a hypha; **-genous,** living inside or undergoing development within; **-parasite,** a parasite living inside its host; **-phyte,** a plant living inside another organism; **-trophic** (of mycorrhizae), having the fungus inside the root

enzyme—protein that catalyzes a specific biochemical reaction

epi- (prefix)—upon; **-biotic,** living on the surface of another; **-cotyl**, portion of an embryo axis above the attachment of the cotyledon(s); the growing apex of the epicotyl forms the young stem of a plant; **-demic** (adj.) general and serious outbreak of disease in man, but used loosely of plants and animals; **-demiology,** study of factors influencing initiation, development, and spread of infectious disease; **-dermis**, surface layer of cells of leaves and other soft plant parts; **-geous,** growing on the surface; **-phyllous,** on the upper surface of a leaf; **-phyte,** a plant living on another but not as a parasite; **-phytotic,** sudden and destructive development of a plant disease over an extensive area; an epidemic

epigynous—having the antheridium above the oogonium on one hypha

eradicate—destroy or remove after disease is established

erumpent—bursting or erupting through the substrate surface
estrogen (adj. **estrogenic**)—substance (as a sex hormone) tending to promote estrus and stimulate the development of secondary sex characteristics in the female
exo- (prefix)—outside; **-genous,** undergoing development outside; **-spore,** the outer layer of a spore wall
exserted—sticking out; protruding
extrude—to push out; emit to the outside
exudate—liquid (ooze) excreted or discharged from diseased tissue or from roots and leaves

facultative parasite—saprophytic organism capable of parasitism
falcate, falciform—curved like the blade of a scythe or sickle
fallow—maintenance of land without plant growth
fasciation—distortion of a plant caused by an injury or infection that results in thin, flattened, and sometimes curved shoots; the plant may look as if several of its stems were fused
fascicle (especially of fungi)—a small group, bundle, or cluster
fecund, fecundity—fruitful in offspring or vegetation; prolific
feeder roots—fine roots with large absorbing area
fibrous—containing, consisting of, or resembling fibers
fibrovascular—having or consisting of fibers and conducting cells (as vessels)
filamentous, filiform—threadlike
fission—form of cell division; splitting into two by division of the complete organism
flaccid—limp or flabby as opposed to turgid; not stiff
flagellum (pl. **flagella**)—hair, whip- or tinsel-like appendage or motile cell (bacteria, zoospores) that provides locomotion; a flagellum is similar to a cilium but is longer
flag leaf—leaf originating from the first culm node below the rachis; uppermost leaf
fleck—small spot; often white to translucent and visible through a leaf
flexuous—having turns, bends, or windings
flexuous (receptive) hypha (of rust fungi, Uredinales)—branched or unbranched, haploid hyphal projection from a pycnium that may be dikaryotized by a pycniospore of the opposite mating type
floret—small flower enclosed in a spikelet
fluorescent—giving out light when placed in ultraviolet (or other) radiation
foliar—pertaining to leaves
foot cell—basal cell as of a spore (e.g., *Fusarium*) or supporting the conidiophore (e.g., *Aspergillus, Sclerospora*)
frass—excrement of an insect
fruiting body, fructification—general term for complex, spore-bearing fungal structure, e.g., apothecium, ascocarp, basidiocarp, perithecium, pycnidium
fuliginous—sooty; obscure; having dark or dusky color
fumigant—vapor-active (volatile) disinfectant used to kill microorganisms and other pests
fungicide (adj. **fungicidal**)—chemical or physical agent that kills or inhibits the growth of fungi
fungi imperfecti (deuteromycetes)—group of fungi for which the sexual stage is not known or the asexual stage of ascomycetes
fungistatic, fungistasis (= *mycostasis*)—chemical or physical agent that inhibits fungal growth or spore germination without killing
fungus (pl. **fungi**)—organism having no chlorophyll, reproduction by sexual or asexual spores and not by fission, and, generally, a mycelium with well-marked nuclei
fusiform—spindlelike; narrowing toward the ends
fusoid—somewhat fusiform

gall—an abnormal swelling or localized outgrowth, often more or less spherical, produced by a plant as the result of attack by a fungus, bacterium, insect, mite, or other agent
gamete—sex cell; especially a cell formed in a gametangium that fuses with another sex cell in sexual reproduction
ganglion (pl. **ganglia**)—mass of nerve tissue containing nerve cells external to the brain or spinal cord
gene—smallest functional unit of genetic material (a hereditary determiner) on a chromosome; unit of inheritance
geniculate—bent, like a knee
genome—set or group of chromosomes
genotype—genetic constitution of an individual or group; class or group of individuals sharing a specific genetic makeup
genus (pl. **genera**)—group of related species
germinate—begin growth as of a seed, spore, sclerotium, or other reproductive body
germ pore—differentiated thin area within a spore wall (especially in rusts) through which a germ tube can emerge
germ tube—hypha resulting from an outgrowth of the spore wall and/or cytoplasm
glob-oid, -ose, -ular, -ulose—almost spherical
Gramineae—taxonomic family name for grasses
gramineous—of or relating to a grass
gram-positive or **-negative**—bacteria staining violet or red, respectively, following treatment with Gram's stain
gregarious—in groups or companies but not joined together
guttation—exudation of watery, sticky liquid from stomates or hydathodes
gyre—a circular or spiral motion or form

haploid—having a single set of chromosomes (n)
haustorium (pl. **haustoria**)—special hyphal branch of a parasite, especially one within a living host cell, for absorption of food. Often associated with rusts, downy and powdery mildews, parasitic flowering plants, and other obligate parasites
helical—of, relating to, or having the form of a helix
helix—something spiral in form
herbaceous—plants that do not develop much woody tissue, thus remaining soft and succulent
herbicide—chemical or physical agent that limits the growth of or kills plants
heter- (prefix)—other; not normal; different; **-oecious** (noun, **-oecism**), undergoing different parasitic stages on two unlike hosts, as in the rust fungi; **-ogamy,** the union of gametes of different form; **-okaryosis** (adj. **-okaryotic**), (of cells), the condition of having two or more genetically different nuclei, sometimes as a result of anastomosis; (of mycelia) being made up of heterokaryotic cells; **-othallic,** having sexes separated in different mycelia; **-otrophic** (of living organisms), using organic compounds as primary sources of energy; **-ozygous,** having heterokaryosis resulting from the fusion of gametes
hilum—mark or scar, especially that on a spore at the point of attachment to a conidiophore or sterigma; scar on a seed coat, marking the place of attachment of the seed stalk to the seed
homo- (prefix)—one and the same; **-thallism** (adj. **-thallic**), condition in which sexual reproduction can occur without the interaction of two differing thalli; both sexes present in the same mycelium
homologous—like in structure
honey dew—ooze or exudate of sweetish liquid
host—living plant attacked by (or harboring) a living parasite and from which the invader is obtaining part or all of its nourishment
host range—various kinds of plants attacked by a given pathogen
hyaline—transparent or nearly so; translucent; frequently used in the sense of colorless
hybrid—offspring of two individuals of different genetic character
hydathode—special gland or pore structure at ends of vascular tissues through which water of guttation can escape
hymenium—spore-bearing (sporogenous) layer of a fungus fruiting body
hyper- (prefix)—above; **-parasite,** a parasite parasitic on another parasite; **-plasia,** overdevelopment of some sort (e.g., swelling, gall, or witches'-broom); **-sensitivity,** increased sensitivity, as in the condition in which there is death of the host tissue at the point of attack by a pathogen, so that infection does not spread, especially of reaction of rusts; **-trophy** (of organs, etc.), the state of having growth greater than normal due to an abnormal increase in the number of cells
hypha (pl. **hyphae**)—tubular filament of a fungal thallus or mycelium; the unit of structure of the fungi
hyphoid—like hyphae; cobwebby

icosahedral—having the form of an icosahedron (a polyhedron with 20 faces or sides)

immune—not affected by or responsive to disease; exempt from disease
imperfect state (or **stage**)—asexual portion of a life cycle of a fungus; asexual spores (such as conidia) or no spores are produced
inclusion body—subcellular structure induced to form in virus-infected cells
incubation period—time between infection by a pathogen and appearance of symptoms
indeterminate—having the edge not well defined, especially of fruiting bodies and leaf spots; (of conidiophores), continuing growth indefinitely
indicator plant—one that reacts to certain viruses, other pathogens, or environmental factors with specific symptoms. Such plants are used for identification of pathogens or environmental factors
indigenous—native; natural to a country or region
infect, infection (of a pathogen)—process of entering (invasion, penetration) and establishing a parasitic relationship with a host plant
infection court—site in or on a host plant where an infection can occur; i.e., root, stalk, leaf, tassel, ear
infection type—gross appearance or size of symptom as from a rust infection
infectious—capable of infecting and spreading from plant to plant
infest—attacked by animals, especially insects or nematodes; to "contaminate" as with microorganisms; to be present in numbers
inflorescence—a flower cluster
injury—result of transient operation of an adverse factor, e.g., insect bite, action of a chemical, or adverse environmental factor
inoculate, inoculation—place inoculum (microorganism or virus, etc.) in an infection court
inoculum—pathogen or its parts, used for inoculating to produce disease
insecticide—physical or chemical agent used to control or kill insects
inter- (prefix)—between, among; **-calary cell,** a cell between two others; **-cellular,** between or among (host) cells; **-node,** area between two adjacent nodes on a culm, stem, or stalk
intra- (prefix)—within, inside; **-cellular,** within or through a cell or cells
invade, invasion—growth or movement of a pathogen into a plant and its establishment in it
in vitro—in glass, artificial environment
isolate, isolation—separated or confined spore or microbial culture; process of getting a fungus, bacterium, or other organism into pure culture; the culture itself

karyogamy—fusion of two sex nuclei after cell division, i.e., after plasmogamy
knot—knoblike overgrowth on roots or stems with an imperfect vascular system

labium (pl. **labia**)—lip
lanceolate—like a lance or long spear
larva (pl. larvae)—juvenile; growth stage between the embryo and adult
latent—present but not manifested or visible
leaching—washing of soluble nutrients down through the soil
lenticel—a pore in the bark of woody stems and other plant parts through which exchange of gases occurs
lesion—well-marked but localized diseased area; a wound
linear—long and narrow
lobate—lobed
locule, loculus—a cavity, especially one in a stroma
lodging—laying down; stalks approach horizontal rather than vertical position
lumen—central cavity of a cell or tubular organ (e.g., stylet of nematodes); hollow center of a culm (grass stem)

macro- (prefix)—long; **-conidium,** the larger, generally more diagnostic, conidium of a fungus (e.g., *Fusarium*) that also has microconidia, a long or large conidium; **-scopic,** visible with the naked eye; **-spore,** a large spore where there are spores of two sizes
malac- or **malaco-** (prefix)—of or relating to soft
male-sterile—having nonfunctional or no male sex organs
malpighian tube—any of a group of long blind vessels opening into the posterior part of the alimentary canal in most insects and functioning primarily as excretory organs
manual transmission (or **inoculation**)—spread or introduction of inoculum to an infection court by hand manipulation
matrix—substratum in or on which a fungus is living
mechanical transmission (or **inoculation**)—spread or introduction of inoculum to an infection court (especially a wound) by hand manipulation accompanied by a physical disruption of the host tissue
meiosis—reduction division; process by means of which chromosome numbers are reduced by half; process by which a zygote (2n) divides and produces haploid (n) gametes
melanin (adj. **melanoid**)—brown-black pigment
meristem (adj. **meristematic**)—plant tissue that functions principally in cell division and differentiation; a mass of growing cells, capable of frequent cell division
mesophyll—chlorophyllous tissues of a leaf between the epidermal layers
metabolism—total of the chemical transformations occurring in the body of a living organism
metabolite—chemical component of a reaction series of process
micro- (prefix)—small; **-conidium,** the smaller conidium of a fungus (e.g., *Fusarium*) that also has macroconidia, a small conidium often acting as a spermatium; **-flora,** composite of microscopic plants; **-scopic,** too small to be seen except with aid of a microscope; **-spore,** a small spore where there are spores of two sizes
midrib—central, thickened vein of grass leaves
mildew (downy and powdery)—plant disease characterized by a thin coating of mycelial growth and spores on the surfaces of infected plant parts
mitosis—process of nuclear division where chromosomes duplicate, divide, and produce daughter nuclei of the same genetic constitution as the parent nucleus
MLO—mycoplasmalike organism
mold—fungus with conspicuous, profuse, or woolly growth (mycelium or spore masses)
mollicute—group of procaryotic microorganisms lacking a cell wall and bounded by flexuous membranes
molt—shed a cuticle or body encasement during a phase of growth
mono- (prefix)—one; **-cotyledon,** plant with one seed leaf; **-culture,** cultivation of a single crop; **-karyon,** individual with one nucleus per cell; **-typic,** having only one representative, as a genus having only one species
monoecious—having male and female sex structures in one thallus
monotrichous—(of bacteria), having one polar flagellum
morphology—study of form and structure of organisms
mosaic—disease symptom characterized by a mottling of the foliage or by variegated patterns of dark and light green to yellow that form a mosaic; caused by disarrangement of unequal development of the chlorophyll content
mottle—disease symptom comprising light and dark areas in an irregular pattern
mucoid—resembling mucus
mucus—slime; bacterial exudate
multi- (prefix)—a great number, many; much; **-genic,** containing or controlled by a number of genes; **-line,** cultivar with a mixed population of plants; **-nucleate,** having more than one nucleus per cell; **-septate,** having a number of septa; **-sporous,** having a number of spores
multiple infection—invaded by more than one parasite
muriform—having cross and longitudinal septa
mutation (n. **mutant**)—abrupt heritable change in an individual
myc-, mycet-, myceto-, myco- (prefix)—fungus
mycelium (pl. **mycelia**)—mass of hyphae constituting the body (thallus) of a fungus
mycoplasma—prokaryotic organisms, smaller than conventional bacteria, without rigid cell walls and variable in shape
mycorrhiza (pl. **mycorrhizae** or **mycorrhizas**)—association or resting structure of a symbiotic, nonpathogenic, or weakly pathogenic fungus and the roots of plants
mycosis (pl. **mycoses**)—fungal disease of man, animals, or, rarely, plants. Mycoses are frequently named after the part attacked (*broncho-* [respiratory tract], *dermato-* [skin], or

pneumo-[lungs])

mycotoxicoses—literally fungus poisonings; current usage limited to poisoning of man and animals by various food and feed products contaminated (and sometimes rendered carcinogenic) by toxin-producing fungi

mycotoxin—toxin produced by a fungus

necrosis (adj. **necrotic**)—death of plant cells, usually resulting in tissue turning dark; commonly a symptom of fungus, nematode, virus, or bacterial infection

nematicide—chemical or physical agent that kills or inhibits nematodes

nematode—small wormlike animal, parasitic in plants or animals, or free living in soil or water

nocturnal—active or occurring in the night

node—joint of a culm or stalk; site of attachment of leaves and spikelets

nomenspecies (n. sp.)—named species; taxonomic category

nonacid-fast—bacterial cells that are decolorized by 25% sulfuric acid after staining with carbol fuchsin

nonpersistent virus—one that remains infectious within its insect vector for only a short time and may be transmitted immediately after acquisition

nymph—juvenile insect that resembles the adult

obligate—necessary, essential; (of a parasite), living only as a parasite and one that has not been cultured on laboratory media

oblong ellipsoid (of spores)—rounded-oblong; having long parallel sides and ends almost hemispherical

olivaceous—of the color olive or olive-green

onchiostyle—stiff, slender, grooved tooth of stubby-root (*Trichodorus*) nematodes

oogonium (pl. **oogonia**)—female sex organ of oomycetes; one-celled and produces one or more eggs (oospheres)

oosphere—female gamete in oomycetes

oospore—thick-walled, resting spore that develops from an oosphere produced by fertilization or parthenogenesis

organelle—delimited membranous structure within a cell having a specialized function

ostiole, ostiolum—pore; opening in the papilla or neck of a perithecium or pycnidium by which spores are freed

ovary—female reproductive structure that contains an ovule or egg; the basal, enlarged portion of a pistil, within which seeds develop

ovate, ovoid—egg-shaped

oxidase—enzyme of qualitative significance in some bacteria

pallid—light-colored; pale

papilla (pl. **papillae**)—small, rounded, blunt, or nipplelike projection

papillate—having a papilla

paraphysis (pl. **paraphyses**)—sterile, upward growing, basally attached hyphal element in a hymenium, especially in the ascomycetes where they are generally clavate or filiform and branched or unbranched

parasexual—recombination of genetic characters without sexual processes

parasite—organism living in or on another living organism (host) and obtaining food from it

parenchyma—physiologically active plant tissue composed of thin-walled cells that often store food and usually retain meristematic potentialities

parthenogenesis (adj. **parthenogenetic**)—development of an egg (female gamete) into a new individual without fertilization by a male sperm (male gamete)

patho- (prefix)—**-genic,** disease-causing, or able to do so; **-genicity**, ability to cause disease or the condition of being pathogenic; **-toxin,** a toxin or poison produced by a pathogen; **-type,** variant of a given pathogen that gives a pathogenic reaction on a given host

pathogen—organism or agent that causes disease in another organism

pathology—study of diseases and their control

PDA—potato-dextrose agar growth medium

pedicel—small stalk; stalk of an individual flower of an inflorescence

peduncle—stalk bearing a flower, flower cluster, or a fructification; stalk, pedicel

perfect state (stage or **phase)**—state of the life-cycle of fungi in which spores (e.g., ascospores and basidiospores) are formed after nuclear fission; sexual; capable of sexual reproduction

pericarp—outer tissue layer of ripened ovary (maize seed)

peridium—wall or limiting membrane of a sporangium or other fruiting body, or of a rust sorus

peripheral (adv. **peripherally**)—of, relating to, or involving the surface of a body

perithecium (pl. **perithecia**)—flask-shaped or subglobose, thin-walled, fungus fruiting body (ascocarp) containing asci and ascospores. Spores are expelled or otherwise released through a pore (ostiole) at the apex or tip

peritrichous—have flagella distributed along a circumference, e.g., around a bacterial cell

persistent—said of circulatory viruses that remain infectious within their insect or other vectors for long periods, and are transmitted via salivary fluids; nondeciduous; (of teliospore pedicels), remaining firmly attached to the spore and not breaking

pest—any organism injuring plants or plant products

pH—measure of acidity. pH 7 is neutral; below pH 7, acidic; above pH 7, alkaline

phasmid—lateral, caudal swelling on a nematode

phial- (prefix)—**-ide** end cell of a conidiophore or a conidiophore of fixed length with one or more open ends through which a basipetal succession of conidia develop; **-ospore,** conidium borne by a phialide without an increase in length of the phialide itself

phloem—food-conducting tissue in plants

photosynthesis—manufacture of carbohydrate food from carbon dioxide and water in the presence of chlorophyll(s), utilizing light energy and releasing oxygen

phyllody—change of floral parts to foliage leaves

physiogenic (physiological) disease—caused by unfavorable environmental factors

physiological form—subspecies group that differs in behavior or other characteristics but not in morphology. *See* physiologic race

physiologic race—subdivision within a species that is alike morphologically but differs from other races in virulence, symptom expression, biochemical and physiological properties, or, to some extent, in host range

phytopathogenic—capable of causing disease in plants

phytopathology—plant pathology; science of plant disease

phytotoxic—harmful to plants

plasmogamy—fusion between two sexual cells

pleomorphic—able to assume various shapes; having more than one independent form or spore-stage in the life cycle; polymorphic

plumule—bud or shoot-forming meristem of an embryo. *See* epicotyl

polar (of bacteria, spores, etc.)—at the ends or poles

pollen—male sex cells produced by anthers of flowering plants

poly- (prefix)—a great number, many; **-hedron,** that which has many sides or faces; **-morphic,** having different forms, pleomorphic; **-ploid,** having more than two sets of chromosomes; **-sporous,** many spored

pore—small opening

predispose—to make prone to infection and disease

primary infection—first infection by a pathogen after going through a resting or dormant period

primary inoculum—inoculum that initiates rather than spreads or magnifies disease

primary root—root that develops directly from the mesocotyl of an embryo rather than from a crown or node

procaryotic (n. **procaryote**)—without internal membrane-bound organelles

promycelium—basidium of rusts and smuts; initial structure produced upon teliospore germination

propagule—any part of an organism capable of initiating independent growth

protectant—agent, usually a chemical, that prevents or inhibits infection

protein—complex, organic nitrogenous substance, built up from amino acids and constituting the major part of the organic materials in living protoplasm

protoplast—plant cell exclusive of its wall; protoplasm

protuberant—to bulge out, prominent

proximal—nearest; next to or nearest the point of attachment or origin

pseudomonad—bacterium of the genus *Pseudomonas*
pseudothecium (pl. **pseudothecia**)—peritheciumlike ascocarp with a dispersed rather than an organized hymenium; having asci in numerous unwalled locules
pulvinate—cushionlike in form
pupate—to be inactive, as with certain insects in a cocoon or case
pustule—small, blisterlike, frequently erumpent, spot or spore mass (sorus)
pycnidiospore—conidium produced within a pycnidium
pycnidium (pl. **pycnidia**)—asexual, flask-shaped, or globose fungus fruiting body containing conidia (pycnidiospores)
pycniospore—haploid (n), sexually derived spore (spermatium) formed in a pycnium
pycnium (pl. **pycnia**)—haploid (n), pycnidiumlike fruiting body or spermogonium produced by rust fungi

quarantine—legislative control of the transport of plants or plant parts to prevent spread of disease, pathogens, or other pests; law
quiescent—dormant; quiet

race—subgroup or biotype within a species or variety distinguished by behavior (differs in virulence, symptom expression, or, to some extent, in host range) but not by morphology. *See* physiologic race
ramify—to branch; to separate into divisions; to split into branches or constituent parts; to send forth branches or extensions
receptive hyphae—specialized hyphae that protrude from a pycnium and function as female gametes
reniform—kidney-shaped
repeating spore—spore that gives rise to the same type of mycelium as that on which it was produced
resistant, resistance—ability of a host plant to overcome completely or to suppress, prevent, or impede the activity of a pathogen

respiration—series of chemical reactions whereby living protoplasm produces energy at the expense of oxygen, carbohydrate, and fat
resting spore—temporarily dormant spore, usually thick-walled, capable of later germination and initiating infection
reticulate—netted; with netlike ridges
rhabdo virus—small RNA animal virus
rhizoid—stiff, rootlike hyphal strand
rhizome—jointed underground stem
rhizosphere—soil microenvironment near a living root. The microflora is frequently richer than that of the soil away from a root
ringspot—disease symptom characterized by yellowish or necrotic rings with green tissue inside the ring, as in some virus diseases
RNA—ribonucleic acid
rogue (roguing)—remove and destroy by hand undesired individual plants on the basis of disease infection, not true-to-type, insect infestation, or other reason
root cap—thimblelike mass of protective tissue at root apex
root hair—threadlike, single-celled outgrowth from root epidermal cell
rosette—disease symptom characterized by short, bunchy growth habit due to subnormal elongation of internodes
rugose—wrinkled
runner hyphae—thickened hyphal strands

saccate—like a bag or sack
saline—of, relating to, or resembling salt; a natural deposit of common or other soluble salt; a *saline solution* is one isotonic with body fluids
sanitation—destruction of infected and infested plants or plant parts
saprophyte, saprobe (adj. **saprophytic**)—organism that feeds on dead organic matter
scald—blanched appearance of plant parts
sclerotium (pl. **sclerotia**)—hard, frequently rounded, and usually darkly pigmented resting body of a fungus composed of a mass of special hyphal cells. The structure may remain dormant for long periods. Sclerotia germinate upon return of favorable conditions to produce a stroma, fruiting body, mycelium, or conidiophores
scorch—"burning" of plant tissue from infection, lack of some element, chemical injury, or weather conditions
secondary infection—infection resulting from the spread of infectious material produced following a primary infection (the first infection by a pathogen after a resting period) or from other secondary infections without an intervening inactive period
secondary inoculum—inoculum resulting from primary infections
secondary root—root developed from a root crown or node as opposed to a primary root
sedentary—remaining in a fixed location
senescence—decline or degeneration as with maturation, age, or disease stress
septate—with cross-walls; having septa
septum (pl. **septa**)—cross-wall in a hypha
serology—study, detection, and identification of antigens, antibodies, and their reactions
serous—(of latex), like serum; watery opalescent
serrate—edges with teeth, like a saw
sessile—having no stem
seta (pl. **setae**)—stiff, hairlike appendage; usually dark and thick-walled
sheath—lower part of grass (maize) leaf that clasps the culm; membranous cover
shoot—stem with its leaves
shothole—disease symptom where small round fragments drop out of leaves making them appear as if riddled by shot
side-dressing—fertilizer applied to the side of a row crop during growth
sign—any indication of disease from direct visibility of the pathogen or its parts (spores, mycelium, exudate, or fruiting bodies)
sorus (pl. **sori**)—compact fruiting structure, especially the spore mass in the rusts and smuts; occasionally a group of fruiting bodies
sp. (pl. **spp.**)—species; one kind of plant or animal life subordinate to a genus but above a race, strain, or variety. A genus name followed by *sp.* means that the particular species is undetermined. *Spp.* following the genus name means that several species are grouped together without being named individually
spermagonium (pl. **spermagonia**)—flask-shaped, walled fungus structure in which spermatia are produced; a pycnium of a rust fungus
spermatium (pl. **spermatia**)—nonmotile, uninucleate haploid (+ or −) "sex" cell (haploid gamete), variously regarded as a pycniospore or microconidium
spicule—copulatory organ of male nematodes
spike—inflorescence with florets or spikelets on an axis
spikelet—spike appendage comprised of glumes and florets; unit of inflorescence in grasses
sporangiophore—differentiated hypha that bears a sporangium
sporangium (pl. **sporangia**)—saclike or flasklike structure whose entire contents are converted into an indefinite number of asexual spores (sporangiospores, zoospores)
spore—one- to many-celled reproductive body in fungi and lower plants that can develop into a new plant
sporidium (pl. **sporidia**)—basidiospore of rust and smut fungi
sporodochium (pl. **sporodochia**)—conidial spore mass supported by a cushion-shaped mass of short conidiophores interwoven on a stroma
sporulate—to produce spores
stellate—pointed as a star
sterigma (pl. **sterigmata**)—small, usually pointed hyphal branch or structure supporting a spore (conidium, basidiospore) or sporangium
sterol—any of various solid cyclic alcohols (as cholesterol) widely distributed in plant and animal lipids
stigma—part of a pistil, usually the apex, that receives pollen and upon which pollen grains germinate
stipe—stalk
stoma (pl. **stomata**)—regulated opening (pore) in the plant epidermis, controlled by guard cells, for passage of gases and water vapor
strain—biotype; race; an organism or group of organisms differing in minor aspects from other organisms of the same species or variety

streak—an elongated lesion with irregular sides
striate—marked with delicate lines, grooves, or ridges
stroma (pl. **stromata**)—compact mass of specialized fungus hyphae (with or without host tissue or substratum), sometimes sclerotiumlike in form, in or on which fruiting bodies and/or spores are produced
stubble—culm (stalk) bases and crowns of harvested maize still rooted in soil
stunted—unthrifty plant reduced in size and vigor due to unfavorable environmental conditions; may be due to a wide range of pathogens or abiotic agents
stylet—stiff, slender, hollow feeding organ of plant-parasitic nematodes
sub- (prefix)—under, below; often approximating the condition qualified, slightly, somewhat; **-digitate,** not quite having fingerlike divisions radiating from a common center; **-globose** or **-spherical,** almost spherical; **-microscopic**, too small to be seen with an ordinary light microscope; **-strate**, substance or object on which a saprophytic organism feeds and develops; **-ulate**, widest at the base, slender and tapering to a point (awl-shaped)
sunscald—plant tissues "burned" or scorched by excess sun and other unfavorable conditions
susceptible—nonimmune; lacking resistance; prone to infection
swarmspore—zoospore
sympodial (of conidiophores)—characterized by continued growth after the main axis has produced a terminal spore by the development of a succession of apices each of which originates below and to one side of the previous axis
symptom—indication of disease by reaction of the host
symptomatology—study of disease symptoms
syn.—synonym
syncytium (pl. **syncytia**)—multinucleate mass of protoplasm resulting from fusion of protoplasts and surrounded by a common cell wall
synergism, synergistic—association of two organisms acting at one time and affecting a host response that one alone could not make
synnema (pl. **synnemata**)—compact or fused, generally upright group of conidiophores bearing conidia at the apex only or on both apex and sides; coremium
systemic—applied to chemicals or pathogens (or single infections) that spread generally throughout the plant body as opposed to remaining localized

teliospore, teleutospore—thick-walled resting spore produced by some fungi, notably rusts and smuts, that germinates to form a basidium
telium—sorus producing teliospores
teratology—study of malformations, monstrosities, or serious deviations from the normal type in organisms
thallus—fungus body
thorax—insect body part between the head and abdomen
tiller—shoot, culm, or stalk arising from a crown bud
tissue—group of cells, usually of similar structure, that perform the same or related functions
tolerant—capable of sustaining disease without serious damage or yield loss
toxicity—capacity of a substance to produce injury
toxin—poison produced by a living organism
translation (adj. **translational**)—uniform motion of a body in a straight line
translocation—movement of water, minerals, food, or pathogens within a plant
translucent—so clear that light rays may pass through
transmission—spread of virus or other pathogen from plant to plant
transpiration—loss of water vapor from aerial parts of plants, chiefly through stomata in the leaves
trilaminar—consisting of three thin parallel layers
triticale—hybrid cereal from a wheat (*Triticum*) and rye (*Secale*) cross
truncate—ending abruptly as though the end was cut off
turbid—cloudy; not clear
turgid, turgidity—swollen or plump as a result of internal water pressure

ultramicroscopic—too small to be seen with an ordinary laboratory (light) microscope
uni- (prefix)—one; **-cellular,** having one cell; **-flagellate,** having one flagellum; **-nucleate**, having one nucleus; **-seriate**, arranged in one series; **-tunicate**, having one definable wall or cover
uredium (pl. **uredia**)—fruiting body (sorus) of rust fungi that produces uredospores
uredospore, urediospore, urediniospore—binucleate, dikaryotic (n+n), asexual, one-celled, repeating spore of rust fungi borne in a uredium

variety—group of closely related plants of common origin within a species that differ from each other in certain minor details such as form, color, flower, and fruit
vascular—pertaining to conductive (xylem and phloem) tissue
vector—insect, mite, animal, man, etc., able to transmit a pathogen (virus, bacterium, fungus, mycoplasma, nematode)
ventral—front or lower surface as opposed to dorsal
vermiform—wormlike
vesicle—thin, bladderlike sac or structure in which zoospores are differentiated or released; also the bulbous head terminating the conidiophore of *Aspergillus*
vesiculiform, vesicular—having form or structure of a vesicle
vestige (adj. **vestigial**)—trace or visible sign left by something vanished or lost; a minute remaining amount; a small and degenerate or imperfectly developed bodily part or organ that remains from one more fully developed in an earlier stage of the individual, in a past generation, or in closely related forms
viable, viability—state of being alive; ability of seeds, fungus spores, or sclerotia, etc., to germinate
vicid—slimy, sticky, mucilaginous, viscous. *See* viscid
vipipary, viviparous—germinating while still attached to the parent; bear live young as opposed to laying eggs
virescence (adj. **virescent**)—state or condition of turning green
virion—complete virus particle
virulence—degree or measure of pathogenicity; capacity to cause disease
virulent—highly pathogenic
viruliferous—virus-carrying; term applied particularly to virus-laden insects
virus—submicroscopic, filterable agents that cause disease and multiply only in living cells
viscid, viscous—sticky, gluey, half-liquid. *See* vicid
volunteer—self-set plant; plant seeded by chance

water-soaked—disease symptom appearing wet, dark; usually sunken and transparent
whorl—three or more leaves, flowers, twigs, hyphae, or other plant parts arranged in a circle and radiating from one point
wilt—lack of freshness or drooping of leaves from lack of water (inadequate water supply or excessive transpiration); a vascular disease that interrupts the normal uptake and distribution of water
witches'-broom—disease symptom with an abnormal, massed, brushlike development of many weak shoots arising at or close to the same point

xylem—water-conducting tissue in plants

zonate—marked with zones

zoospore—asexually produced fungus spore having cilia or flagella and capable of locomotion in water; a swarm spore

Index